Formal Approaches to Computing
and Information Technology

Also in this series:

Proof in VDM: a Practitioner's Guide
J.C. Bicarregui, J.S. Fitzgerald, P.A. Lindsay, R. Moore and B. Ritchie
ISBN 3-540-19813-X

Systems, Models and Measures
A. Kaposi and M. Myers
ISBN 3-540-19753-2

On the Refinement Calculus

Edited by
Carroll Morgan and Trevor Vickers

Carroll Morgan
Paul Gardiner
Ken Robinson
Trevor Vickers

Springer-Verlag
London   Berlin   Heidelberg   New York
Paris   Tokyo   Hong Kong
Barcelona   Budapest

Carroll Morgan, BSc, PhD
Oxford University Computing Laboratory
Programming Research Group
8–11 Keble Road, Oxford OX1 3QD, UK

Trevor Vickers, BSc, PhD
Department of Computer Science
Australian National University
P.O. Box 4, Canberra 2601, Australia

Series Editor

Steve A. Schuman, BSc, DEA, CEng
Department of Mathematical and Computing Sciences
University of Surrey, Guildford, Surrey GU2 5XH, UK

ISBN 3-540-19809-1 Springer-Verlag Berlin Heidelberg New York
ISBN 0-387-19809-1 Springer-Verlag New York Berlin Heidelberg

British Library Cataloguing in Publication Data
On the Refinement Calculus – (Formal Approaches to Computing &
Information Technology Series)
  I. Morgan, Carroll II. Vickers, Trevor III. Series
  005.1
  ISBN 3-540-19809-1

Apart from any fair dealing for the purposes of research or private
study, or criticism or review, as permitted under the Copyright, Designs
and Patents Act 1988, this publication may only be reproduced, stored
or transmitted, in any form or by any means, with the prior permission
in writing of the publishers, or in the case of reprographic reproduction
in accordance with the terms of licences issued by the Copyright
Licensing Agency. Enquiries concerning reproduction outside those
terms should be sent to the publishers.

© 1992 Carroll Morgan and Trevor Vickers, except where indicated
otherwise for individual articles.
Published by Springer-Verlag London Limited 1994

The use of registered names, trademarks etc. in this publication does
not imply, even in the absence of a specific statement, that such names
are exempt from the relevant laws and regulations and therefore free for
general use.

The publisher makes no representation, express or implied, with regard
to the accuracy of the information contained in this book and cannot
accept any legal responsibility or liability for any errors or omissions
that may be made.

Typesetting: camera-ready by editors
Printed by Antony Rowe Ltd, Bumper's Farm, Chippenham, Wiltshire
34/3830-543210 Printed on acid-free paper

Contents

Introduction ix

The Specification Statement
Carroll Morgan 1

  1 Introduction 1
  2 Specification statements 3
  3 The implementation ordering 7
  4 Suitability of the definitions 8
  5 Using specification statements 10
  6 Miracles 12
  7 Guarded commands are miracles 14
  8 Positive applications of miracles 16
  9 Conclusion 19
10 Acknowledgements 20

Specification Statements and Refinement
Carroll Morgan and Ken Robinson 23

  1 Introduction 23
  2 The refinement theorems 31
  3 The refinement calculus 32
  4 An example: square root 37
  5 Derivation of laws 41
  6 Conclusion 44
  7 Acknowledgements 45

**Procedures, Parameters, and Abstraction:
Separate Concerns**
Carroll Morgan 47

  1 Introduction 47
  2 Procedure call 48
  3 Procedural abstraction 49
  4 Parameters 51
  5 Conclusion 58
  6 Acknowledgements 58

Data Refinement by Miracles
Carroll Morgan 59

1 Introduction 59
2 An abstract program 60
3 A difficult data refinement 61
4 Miraculous programs 61
5 Eliminating miracles 62
6 Conclusion 63
7 Acknowledgements 64

Auxiliary Variables in Data Refinement
Carroll Morgan 65

1 Introduction 65
2 The direct technique 66
3 The auxiliary variable technique 66
4 The correspondence 67
5 Conclusion 69
6 Acknowledgements 70

Data Refinement of Predicate Transformers
Paul Gardiner and Carroll Morgan 71

1 Introduction 71
2 Predicate transformers 72
3 Algorithmic refinement of predicate transformers 74
4 Data refinement of predicate transformers 74
5 The programming language 76
6 Distribution of data refinement 79
7 Data refinement of specifications 81
8 Data refinement in practice 82
9 Conclusions 83
10 Acknowledgements 84

Data Refinement by Calculation
Carroll Morgan and Paul Gardiner 85

1 Introduction 85
2 Refinement 86
3 Language extensions 90
4 Data refinement calculators 92
5 Example of refinement: the "mean" module 96
6 Specialized techniques 101
7 Conclusions 107
8 Acknowledgements 108
9 Appendix: refinement laws 108

A Single Complete Rule for Data Refinement
Paul Gardiner and Carroll Morgan 111

1 Introduction 111
2 Data refinement 112
3 Predicate transformers 115
4 Completeness 116
5 Soundness 118
6 Partial programs 120
7 An example 123
8 Conclusion 125
9 Acknowledgements 126

Types and Invariants in the Refinement Calculus
Carroll Morgan and Trevor Vickers 127

1 Introduction 127
2 Invariant semantics 128
3 The refinement calculus 129
4 A development method 133
5 Laws for local invariants 135
6 Eliminating local invariants 137
7 Type-checking 139
8 Recursion 141
9 Examples 143
10 A discussion of motives 150
11 Related work 151
12 Conclusions 152
    Acknowledgements 153
A Additional refinement laws 153

References 155

Authors' Addresses 159

Introduction

The refinement calculus is a notation and set of rules for deriving imperative programs from their specifications. It is distinguished from earlier methods (though based on them) because the derivations are carried out within a single 'programming' language: there is no separate language of specifications.

That does not mean that specifications are executable; it means rather that "not all programs are executable" [3]. Some are written too abstractly for any computer to execute, and they are the opposite extreme to those which — though executable — are too complex for any human to understand. *Program derivation* is the activity that transforms one into the other.

R.-J.R. Back [4] first extended Dijkstra's language of guarded commands with specifications, and has more recently developed it further [5, 6, 7]. J. Morris also does significant research in this area [50, 51]. This collection however includes only the work done at Oxford.

The refinement calculus is distinguished from some other 'wide spectrum' approaches (*e.g.*, [13, 24]) by its origins and scale: it is a simple programming language to which specifications have been added. The extension is modest and unprejudiced, and one can sit back to see where it leads. So far, it has uncovered 'miracles' [44, 50, 42], novel techniques of data refinement [4, 51, 20, 47, 19], a simpler treatment of procedures [43, 6], 'conjunction' of programs [40, 20, 47], and a light-weight treatment of types in simple imperative programs [45]. It has been applied by Back and his colleagues to parallel and reactive systems [10, 9], and some work has begun on designing a refinement calculus of expressions (rather than of state-transforming programs), by Gardiner and Martin [38] and, independently, by Morris [52].

The work at Oxford sprang from the need to develop a rigorous imperative programming course for our first undergraduate computing degree, which began only in 1985. The strong local tradition of specification and refinement, in Z [25, 49, 59], had not yet reached the level of everyday programming languages (a disadvantage with respect to VDM, for example); the *specification statement*, like a Z schema but with *wp*-semantics, was to make the connection. The current text for that course [46] rests on the subsequent research, reported in this volume, but has been much informed by the work of others. As an undergraduate *programming* text, it presents its method mostly by example, and is deliberately light on justification and history. Thus those seeking the background and mathematical underpinnings should look in the present collection.

The specification statement introduces specifications to Dijkstra's language of guarded commands, and explores the consequences: increased expressive power, the new prominence of the refinement relation, miracles, and a surprising factorisation of that language into smaller pieces. It is the same factorisation reported independently by Nelson [54].

Specification statements and refinement gives our first collection of 'laws of refinement', and lays the emphasis on a *calculus* of refinement.

Procedures, parameters, and abstraction: separate concerns shows how specifications in a programming language allow the *copy rule* of ALGOL 60, once again, to give the meaning of procedures. One side effect is the ability to parametrize program fragments which are *not* procedures.

Data refinement using miracles and *Auxiliary variables in data refinement* describe small aspects of data refinement, independently of the refinement calculus. The former uses the Gries and Prins data-refinement rule [23] only in order to be self-contained. Data refinement is dealt with more generally in *Data refinement of predicate transformers* and *Data refinement by calculation*. The first gives a more theoretical, the second a more practical exposition of the way data refinement and the refinement calculus can interact.

Types and invariants in the refinement calculus explains the way in which the hitherto informal treatment of types during *wp*-based program derivation can be made rigorous without undue formal cost. It exposes some surprising connections — for example that ascribing a type to a variable is a degenerate form of data refinement. It leads also to some unexpected conclusions, one of which is that 'ill-typed' programs can be considered well-formed but miraculous.

A single complete rule for data refinement brings two separate techniques of data refinement, known to be jointly-complete up to bounded nondeterminism, into a single method. The separate techniques, based on relations, are able to be integrated only once they have been reformulated in terms of predicate transformers. Work remains to be done on the more calculational aspects of the general rule.

A reasonable overview can be gained by reading *Specification statements and refinement* and *Data refinement by calculation*.

There is some overlap between the papers: the introduction to *Specification statements and refinement* repeats material from *The specification statement; Auxiliary variables in data refinement* amplifies a section of *Data refinement by calculation*; and various laws of program refinement appear in two places: in *Specification statements and refinement* and as an appendix to *Data refinement by calculation*. (A more comprehensive collection is given in [46].)

September 1992   Carroll Morgan
                                                        Trevor Vickers

The Specification Statement

Carroll Morgan

Abstract

Dijkstra's programming language is extended by *specification statements*, which specify parts of a program "yet to be developed." A weakest precondition semantics is given for these statements, so that the extended language has a meaning as precise as the original.

The goal is to improve the *development* of programs, making it more as it should be: manipulations within a single calculus. The extension does this by providing one semantic framework for specifications and programs alike — developments begin with a program (a single specification statement), and end with a program (in the executable language). And the notion of *refinement* or *satisfaction*, which normally relates a specification to its possible implementations, is automatically generalised to act between specifications and between programs as well.

A surprising consequence of the extension is the appearance of *miracles:* program fragments that do not satisfy Dijkstra's *Law of the Excluded Miracle*. Uses for them are suggested.

1 Introduction

Dijkstra in [17] introduces the *weakest precondition* of a program P with respect to a postcondition *post*; following [27] we will write this $P < post >$. In this style, a *specification* of a program P is written

$pre \Rightarrow P < post >$.

This means "if activated in a state for which *pre* holds, the program P must terminate in a state for which *post* holds."

In traditional top-down developments, we build algorithmic structure around a collection of ever-decreasing program fragments "yet to be implemented," and at any stage we have specifications for those fragments. Thus one finds the dictions

$$\vdots \\ P; \\ \vdots$$

where $pre \Rightarrow P < post >$.

The letter P stands for the missing fragment, and the **where** clause gives its specification. But in our approach, we write instead

$$\vdots \\ [pre, post]; \\ \vdots \tag{1}$$

We write the specification itself at the point to be occupied by its implementation. More significantly, by giving a weakest precondition semantics to $[pre, post]$, we make this intermediate stage (1) into a *program* — albeit an abstract one.

Program development we see as analogous to solving equations: one transforms an abstract program into a concrete one, just as one transforms a complex equation (*e.g.*, $x^2 - x - 1 = 0$) into a simple equality (*e.g.*, $x = (1+\sqrt{5})/2$). For such formulæ, the manipulations are mediated by the relation of implication: the simple equality *implies* the complex equation.

The abstract-to-concrete transformation of programs is mediated by a relation $\sqsubseteq$ of *refinement*, which is defined so that $P \sqsubseteq Q$ means "any specification satisfied by P is satisfied by Q also." This relation can appear between abstract programs (specifications), between concrete programs, or between one and the other. As we write

$$x^2 - x - 1 = 0 \quad \Leftarrow \quad x = \tfrac{1+\sqrt{5}}{2},$$

so we will write with complete rigor

$$x\colon \left[x^2 - x - 1 = 0\right] \quad \sqsubseteq \quad x := \tfrac{1+\sqrt{5}}{2}.$$

An unexpected consequence of our extension is the introduction of abstract programs that do not obey Dijkstra's *Law of the excluded miracle*. These correspond to specifications that have no concrete solution, just as negative numbers stand for insoluble equations in elementary arithmetic ("3 from 2 won't go"). An example is the statement $[true, false]$; we will see that the following holds:

$$true \Rightarrow [true, false] < false > .$$

But just as negative numbers simplify arithmetic, miracles simplify program derivation.

Our overall contribution is *uniformity*: we place program development within reach — in principle — of a single calculus. We expect this to be useful not only at the level of small intricacies, but in the larger scale also. Modules, for example, can be written using specification statements instead of concrete constructions: thus we have *specifications* of modules. Because of the generality of our approach, any structuring facility offered by the target programming language is offered to specifications also.

2 Specification statements

We introduce the syntax and weakest precondition semantics of specification statements, moving from simple to more general forms.

2.1 The simple form

The simple specification statement $[pre, post]$ comprises two predicates over the program variables $\vec{v}$. Informally, it means "assuming an initial state satisfying *pre*, establish a final state satisfying *post*." Its precise definition is (using $\hat{=}$ for "is defined to be")

Definition 1 $[pre, post] < R > \quad \hat{=} \quad pre \land (\forall \vec{v}. \; post \Rightarrow R)$ ♡

For example, assuming $\vec{v}$ is just the single variable x, we have

$$\begin{aligned}
& [true, x = 1] < R > \\
= \; & true \land (\forall x. \; x = 1 \Rightarrow R) \\
= \; & R[x \backslash 1].
\end{aligned}$$

The substitution $[x \backslash 1]$ denotes syntactic replacement of x by 1 in the usual way.

2.2 Confining change

We allow the changing of variables to be confined to those of interest. For any subvector $\vec{w}$ of $\vec{v}$, the statement $\vec{w} : [pre, post]$ has the following informal meaning:

> assuming an initial state satisfying *pre*, establish a final state satisfying *post* while changing only variables in $\vec{w}$.

The precise definition of $\vec{w} : [pre, post]$ is

Definition 2 $\vec{w} : [pre, post] < R > \quad \hat{=} \quad pre \land (\forall \vec{w}.\ post \Rightarrow R)$ ♡

The only change from definition 1 is that the vector of quantified variables is now $\vec{w}$ rather than $\vec{v}$. Taking $\vec{v}$ to be "x, y", we have

$$\begin{aligned} & x : [true, x = y] < R > \\ = \ & true \land (\forall x.\ x = y \Rightarrow R) \\ = \ & R[x \backslash y]. \end{aligned}$$

Since $(x := y) < R >$ equals $R[x \backslash y]$ also, we have shown that $x : [true, x = y]$ and $x := y$ have the same meaning. If we allow *both* x and y to change, this is no longer true:

$$\begin{aligned} & x, y : [true, x = y] < R > \\ = \ & true \land (\forall x, y.\ x = y \Rightarrow R) \\ = \ & (\forall y.\ R[x \backslash y]). \end{aligned}$$

The statement $x, y : [true, x = y]$ can set y to x, x to y, or both x and y to some third value.

2.3 Referring to the initial state

Occurrences of 0-subscripted variables $\vec{v}_0$ in *post* refer to the values held by those variables *initially*. We reserve 0-subscripts for this purpose, and assume that they do not occur as ordinary variables in programs. We now have the following informal meaning for $\vec{w} : [pre, post]$:

> assuming an initial state satisfying *pre*, change only variables in $\vec{w}$ to establish *post*, in which 0-subscripted variables refer to the values those variables held initially.

The precise definition appears below. In practice, however, we usually apply the simpler version given in lemma 1 following.

Definition 3

$$\vec{w} : [pre, post] < R > \;\hat{=}\; pre \wedge (\forall\, \vec{w}.\; post[\vec{v}_0\backslash\vec{f}\,] \Rightarrow R)[\vec{f}\,\backslash\vec{v}\,]$$

where $\vec{f}$ is some fresh vector of variables.

♡

The use of fresh variables $\vec{f}$ in definition 3 is only to avoid interference with possible occurrences of $\vec{v}_0$ in R, which are rare in practice. Usually we can apply the simpler construction below:

Lemma 1 *If R contains no 0-subscripted variables,*

$$\vec{w} : [pre, post] < R > \;=\; pre \wedge (\forall\, \vec{w}.\; post \Rightarrow R)[\vec{v}_0\backslash\vec{v}\,]$$

Proof: Immediate from definition 3.

♡

Notice that if *post* contains no $\vec{v}_0$, then both definition 3 and lemma 1 reduce to definition 2.

For example, taking $\vec{v}$ to be "x, y" as before we have from lemma 1

$$\begin{aligned}
&\; x : [true, x = x_0 + y_0] < R > \\
=&\; true \wedge (\forall\, x.\; x = x_0 + y_0 \Rightarrow R)[x_0, y_0\backslash x, y] \\
=&\; R[x\backslash x_0 + y_0][x_0, y_0\backslash x, y] \\
=&\; R[x\backslash x + y].
\end{aligned}$$

2.4 The implicit precondition

We allow the omission of the precondition in a specification statement. The informal meaning of $\vec{w} : [post]$ is

> assuming it is possible to do so, change only variables in $\vec{w}$ to establish *post*, in which 0-subscripted variables refer to the values those variables held initially.

The meaning is given syntactically — we make the missing precondition explicit:

Definition 4 $\vec{w} : [post] \;\widehat{=}\; \vec{w} : [\,((\exists\,\vec{w} \bullet post))[\vec{v}_0\backslash\vec{v}], \; post\,]$ ♡

For example, we can write

$$\begin{array}{ll} m: [l \leq m \leq h] & \text{for} \\ \text{and} \quad i: [a[i] = v] & \text{for} \end{array} \quad \begin{array}{l} m: [l \leq h,\; l \leq m \leq h] \\ i: [((\exists\,i \bullet a)\overline{[i]} = \overline{v}),\; a[i] = v] \end{array}$$

The first statement places m between l and h; the second locates an index i of value v in array a. If in either case the result is not achievable (*e.g.*, if l exceeds h, or v does not occur in a), the statement can abort.

2.5 Generalised assignment

We generalise assignment by giving the following meaning to the statement $x :\odot e$, for any binary relation $\odot$:

> assuming it is possible to do so, assign to x a value bearing the relation $\odot$ to the expression e, where occurrences of x in e refer to its initial value.

Ordinary assignment statements are now the special case in which $\odot$ is "=". But we can also write, for example,

$$\begin{array}{ll} x :\in s & \text{for} \\ \text{and} \quad n :< n & \text{for} \end{array} \quad \begin{array}{l} \textit{if possible, choose } x \textit{ from } s \\ \textit{decrease } n. \end{array}$$

The definition is given syntactically:

Definition 5 $x :\odot e \;\widehat{=}\; x : [x \odot e[x\backslash x_0]]$ ♡

With this definition, our abbreviations above become respectively

$$\begin{array}{ll} x : [x \in s] & \text{(that is, } x : [s \neq \{\}, x \in s]) \\ \text{and} \quad n : [n < n_0]. & \end{array}$$

The syntax for generalised assignment was suggested (long ago) by Jean-Raymond Abrial.

3 The implementation ordering

For programs P and Q, we give $P \sqsubseteq Q$ the informal meaning: "every specification satisfied by P is satisfied by Q also." This means that Q is an acceptable replacement for P. Our precise definition is

Definition 6 $P \sqsubseteq Q$ *iff for all predicates R,*

$$P < R > \Rightarrow Q < R >.$$

♡

The following theorem shows definition 6 to have the property we require:

Theorem 1 *If* $pre \Rightarrow P < post >$ *and* $P \sqsubseteq Q$, *then also*

$$pre \Rightarrow Q < post >.$$

Proof: Since $P \sqsubseteq Q$, we have $P < post > \Rightarrow Q < post >$. The result follows immediately.
♡

As an example of refinement between *programs*, let P be

if $2|x \rightarrow x := x \div 2$
[] $\ 3|x \rightarrow x := x \div 3$
fi,

and let Q be

if $2|x \ \ \ \ \rightarrow x := x \div 2$
[] $\ \neg(2|x) \rightarrow x := x \div 3$
fi,

where $2|x$ means "2 divides x exactly", and $\div$ denotes integer division. We have $P \sqsubseteq Q$ because

$$P < R > \quad = \quad \begin{array}{l} (2|x \vee 3|x) \hspace{2.5em} \wedge \\ (2|x \ \Rightarrow R[x \backslash x \div 2]) \ \wedge \\ (3|x \ \Rightarrow R[x \backslash x \div 3]) \end{array}$$

and

$$Q < R > \quad = \quad \begin{array}{l}(2|x \quad \Rightarrow R[x\backslash x \div 2]) \quad \wedge \\ (\neg(2|x) \quad \Rightarrow R[x\backslash x \div 3]).\end{array}$$

Thus $P < R > \Rightarrow Q < R >$ for any R. But Q differs from P in that Q will always terminate, even when $x = 7$. And Q is deterministic: if $x = 6$, Q will establish $x = 3$. In spite of these differences, Q is an acceptable substitute for P, and that is why we can implement P as IF $2|x$ THEN $x := x \div 2$ ELSE $x := x \div 3$ END.

We now state the well-known but crucial fact that the program constructors are monotonic with respect to $\sqsubseteq$; only this ensures that refining a fragment (say P above) "in place," in some larger program, refines that larger program overall.

Theorem 2 *If $F(P)$ is a program containing the program fragment P, and for another program fragment Q we have $P \sqsubseteq Q$, then*

$$F(P) \sqsubseteq F(Q)$$

Proof: Structural induction, over the program constructors ";", "**if**", and "**do**".
♡

4 Suitability of the definitions

We now show the suitability of our definitions by proving that

$$pre \Rightarrow P < post > \quad \text{iff} \quad [pre, post] \sqsubseteq P.$$

In fact, we prove a stronger result, dealing with the general form of section 2.3.

In long formulæ, we will sometimes "stack" conjunctions for clarity, writing

$$\begin{pmatrix} this \\ that \end{pmatrix} \quad \text{for} \quad (this \wedge that).$$

Our theorem is a consequence of the following two lemmas.

Lemma 2 *If $\vec{u}$ and $\vec{w}$ partition the vector $\vec{v}$ of program variables, then*

$$pre \wedge \vec{v} = \vec{v}_0 \quad \Longrightarrow \quad \vec{w} : [pre, post] < post \wedge \vec{u} = \vec{u}_0 >$$

Proof: Here we must use definition 3 rather than lemma 1, since the post-condition contains $\vec{v}_0$. We have

$$(pre \wedge \vec{v} = \vec{v}_0) \Longrightarrow \vec{w} : [pre, post] < post \wedge \vec{u} = \vec{u}_0 >$$

if   by definition 3,

$$(pre \wedge \vec{v} = \vec{v}_0) \Longrightarrow pre \wedge \left(\forall \vec{w}.\ post[\vec{v}_0 \backslash \vec{f}] \Rightarrow \begin{array}{c} post \\ \vec{u} = \vec{u}_0 \end{array} \right) [\vec{f} \backslash \vec{v}]$$

if $\vec{v} = \vec{v}_0 \Longrightarrow \left(\forall \vec{w}.\ post[\vec{v}_0 \backslash \vec{f}] \Rightarrow \begin{array}{c} post \\ \vec{u} = \vec{u}_0 \end{array} \right) [\vec{f} \backslash \vec{v}]$

if $\vec{v} = \vec{v}_0 \Longrightarrow \left(\forall \vec{w}.\ post[\vec{v}_0 \backslash \vec{f}] \Rightarrow \begin{array}{c} post \\ \vec{u} = \vec{u}_0 \end{array} \right) [\vec{f} \backslash \vec{v}_0]$

if $\vec{v} = \vec{v}_0 \Longrightarrow \left(\forall \vec{w}.\ post \Rightarrow \begin{array}{c} post \\ \vec{u} = \vec{u}_0 \end{array} \right)$

if   since $\vec{u}, \vec{w}$ partition $\vec{v}$
       true.

♡

Lemma 3 *If* $pre \wedge \vec{v} = \vec{v}_0 \Longrightarrow P < post \wedge \vec{u} = \vec{u}_0 >$ *then*

$$\vec{w} : [pre, post] \sqsubseteq P$$

where $\vec{w}$ and $\vec{u}$ partition the program variables $\vec{v}$.

Proof:

$$pre \wedge \vec{v} = \vec{v}_0 \Longrightarrow P < post \wedge \vec{u} = \vec{u}_0 >$$

hence   by distributivity of $\Rightarrow$ over weakest preconditions,

$$pre \wedge \vec{v} = \vec{v}_0 \wedge \left(\forall \vec{v}.\ \begin{array}{c} post \\ \vec{u} = \vec{u}_0 \end{array} \Rightarrow R \right)$$
$$\Longrightarrow P < R >$$

hence $pre \wedge \vec{v} = \vec{v}_0 \wedge (\forall \vec{w}.\ post \Rightarrow R)[\vec{u} \backslash \vec{u}_0] \Longrightarrow P < R >$
hence $pre \wedge \vec{v} = \vec{v}_0 \wedge (\forall \vec{w}.\ post \Rightarrow R) \Longrightarrow P < R >$

hence   since pre and $P < R >$ do not contain $\vec{v}_0$,
           $pre \wedge (\forall \vec{w}.\ post \Rightarrow R)[\vec{v}_0 \backslash \vec{v}] \Longrightarrow P < R >$

hence   by lemma 1,
           $\vec{w} : [pre, post] < R > \Longrightarrow P < R >$.

Since R was arbitrary, we conclude from definition 6 that $\vec{w} : [pre, post] \sqsubseteq P$ as required.
♡

Those two lemmas give us our theorem immediately:

Theorem 3 *If $\vec{w}$, $\vec{u}$ partition the program variables $\vec{v}$, then*

$$pre \wedge \vec{v} = \vec{v}_0 \implies P < post \wedge \vec{u} = \vec{u}_0 >$$

if and only if

$$\vec{w} : [pre, post] \sqsubseteq P.$$

Proof: "If" follows from lemma 2 and theorem 1; "only if" is lemma 3 exactly.
♡

5 Using specification statements

For illustration we take the simplest of examples: suppose we are given an array $a[0..N-1]$ and must find an index i at which the value v occurs. And we may assume there is such an i. The program is

$$i: \left[\begin{array}{c} 0 \leq i < N \\ \overline{a[i] = v} \end{array} \right] \qquad (2)$$

This *is* a program, though abstract, and perhaps we can execute it directly (see further below). But for now, we assume not — and so we "solve" it, refining it to statements we can execute.

First we use definition 4, rewriting

$$i: \left[(\exists\, i.\, 0 \leq i < N \wedge a[i] = v) \,,\, \begin{array}{c} 0 \leq i < N \\ \overline{a[i] = v} \end{array} \right] \qquad (3)$$

We take as invariant

$$Inv \ \widehat{=}\ \begin{array}{c} 0 \leq i < N \\ (\exists\, j.\, i \leq j < N \wedge a[j] = v) \end{array}$$

The variant is $N - i$. With these and theorem 3, we can prove that (3)

$$\sqsubseteq \quad i: [(\exists\, i.\, 0 \leq i < N \wedge a[i] = v)\,,\ Inv];$$
$$\text{do } a[i] \neq v \rightarrow$$
$$\quad i: \left[\begin{array}{cc} Inv & i_0 < i \\ a[i] \neq v & Inv \end{array}\right]$$
$$\text{od}$$

Notice that the fragments "to be developed" are written in-line, and that the above mixture of abstract and concrete is still a program. The first component we can refine to $i := 0$; and the second we can refine to $i := i + 1$. For illustration, we show the second refinement in more detail: by theorem 3 we need

$$\left(\begin{array}{c} 0 \leq i < N \\ (\exists j.\, i \leq j < N \wedge a[j] = v) \\ a[i] \neq v \\ i = i_0 \end{array}\right) \Longrightarrow i := i + 1 \left\langle \begin{array}{c} i_0 < i < N \\ (\exists j.\, i \leq j < N \wedge a[j] = v) \end{array} \right\rangle$$

By the semantics of assignment [17], the consequent is

$$i_0 < i + 1 < N$$
$$(\exists j.\, i + 1 \leq j < N \wedge a[j] = v).$$

That follows easily from the antecedent.

Having our development, we may wish to collect it and others into a small "database module," based on arrays. As is typical in modern programming languages, the implementation

$$i := 0;$$
$$\text{do } a[i] \neq v \rightarrow i := i + 1 \text{ od}$$

would be hidden within the "implementation part" of the module. What should appear in the definition part? We suggest (using the syntax of Modula-2 [62])

```
module Database;
    export Find, N;
    const N = ?;
    var a: array [0..N − 1] of ?;
    procedure Find(v: ?; var i: [0..N − 1]);
    begin
        i: [ 0 ≤ i < N
             a[i] = v  ]
    end Find
    ⋮
end Database
```

This is not informal. Except for the "?," the module contains only constructions whose semantics are known precisely. Now a programmer wishing to implement (2) can do so directly, using the copy rule of Algol-60 (suitably extended for modules). He just writes $Find(v, i)$, whose meaning is given by substituting the procedure body from the *definition* module. This is discussed further in [43].

Thus we show that our approach applies not only to small constructions, and in particular that it supports the view that the "definition module" specifies the "implementation module."

6 Miracles

In [17] it is stated that for all programs P,

$$P < false >= false. \tag{4}$$

This is no longer true: we have for example

$$\begin{aligned}
& [true, false] < false > \\
=\ & true \wedge (\forall\, \vec{v}.\ false \Rightarrow false) \\
=\ & true.
\end{aligned}$$

The statement $[true, false]$ is called a *miracle*, because it implements anything: we have for all R that $P < R > \Rightarrow [true, false] < R >$, and so for any P whatsoever,

$$P \sqsubseteq [true, false].$$

Although $[true, false]$ implements anything, it cannot itself be implemented by anything free of miracles. This is because "P is free of miracles" implies by (4) that $P < false >= false$, and so taking $R = false$ in definition 6, we have $[true, false] \not\sqsubseteq P$.

A program which cannot be rid of miracles is *infeasible* in the following precise way:

Definition 7 *We say that a program P is* feasible *iff*

$$P < false >= false.$$

Otherwise it is infeasible, *or* miraculous. ♡

Clearly, all programs free of specification statements are by (4) feasible: indeed, they are "implementations" already. For specifications, however, we have the following

Theorem 4 $\vec{w} : [pre, post]$ *is feasible iff*

$$pre \Rightarrow ((\exists \vec{w} \bullet post))[\vec{v}_0 \backslash \vec{v}].$$

Proof: Definitions 3, 7, and predicate calculus.
♡

Miracles can arise "accidentally" in program development if we make an incorrect design step; this is discussed in more detail in [31] and [48]. For the present, we take a trivial example: we (mistakenly) want to implement $x : [x = 0]$ as a sequential composition whose second component is $x := y$. That is, we want to solve the following formula for P:

$$x : [x = 0] \quad \sqsubseteq \quad P; x := y \tag{5}$$

By theorem 3, we have (5)

    iff $x = x_0 \land y = y_0 \Rightarrow (P; x := y) < x = 0 \land y = y_0 >$

    iff   by sequential composition
         $x = x_0 \land y = y_0 \Rightarrow P < x := y < x = 0 \land y = y_0 >>$

    iff $x = x_0 \land y = y_0 \Rightarrow P < y = y_0 = 0 >$

    iff   by theorem 3 again
         $x : [true, y = 0] \quad \sqsubseteq \quad P$

We have found our solution P, showing unconditionally that

$$x\colon [x=0] \;\sqsubseteq\; x\colon [true,\; y=0];\; x:=y$$

In fact, the above shows that $x:[true, y=0]$ is the most general solution of (5), and so we take it as representative of them all, calling it "the" solution. This development technique, in which formulae like (5) are so solved, is the subject of [31].

But, after all, the statement $x\colon [true,\; y=0]$ is infeasible; and the importance of the example is its illustration of that consequence of mistaken design steps. The formula (5) is *not* insoluble, but we cannot develop executable code from its solution.

7 Guarded commands are miracles

Miracles are a strict extension of our programming capabilities — clearly, since they cannot be executed. We now show how close miracles are, nevertheless, to being in the original language.

A guarded command has the syntactic form

$$B \to P,$$

where B is a boolean expression and P is the command guarded. Originally, these occurred only within **if** and **do** constructions. Here we give meaning to guarded commands standing alone.

Informally, we say that a "naked" guarded command *cannot* be executed unless its guard is true. More precisely, we have

Definition 8 $(B \to P) < R > \;\;\hat{=}\;\; B \Rightarrow P < R >$ ♡

If B is true, then $B \to P$ behaves like P. But if B is false, we consider $B \to P$ to be miraculous: we may as well, since in this case we *cannot* execute it to check.

Thus we have a compact notation for miracles: they are naked guarded commands whose guards are not identically true. For example, our first miracle $[true, false]$ can be written for *any* program P

$$false \to P.$$

The following theorem shows that in fact every miracle can be written this way. We have

Theorem 5 *For any program P, feasible or not, there is a guard B and a feasible program Q such that*

$$P = B \rightarrow Q$$

Proof: We take

$$\begin{aligned} B &= \neg P < false > \\ Q &= \textbf{if } B \rightarrow P \textbf{ fi}. \end{aligned}$$

Definition 7 shows that Q is feasible, and definition 8 shows that the equality holds.
♡

We can also define also a non-deterministic composition [] and a "guardless if," achieving correspondence with the original meaning of these constructs. We have

Definition 9 *For any programs P and Q, the program $P \,[]\, Q$ is defined*

$$(P \,[]\, Q) < R > \;\;\widehat{=}\;\; P < R > \wedge Q < R >.$$

♡

Definition 10 *For any program P, the program* **if** P **fi** *is defined*

$$\textbf{if } P \textbf{ fi} < R > \;\;\widehat{=}\;\; \neg P < false > \wedge P < R >.$$

♡

Definition 9 is simple non-deterministic choice; in fact

$$P \,[]\, Q \;=\; \textbf{if } true \rightarrow P \,[]\, true \rightarrow Q \textbf{ fi}.$$

Definition 10 *is* an extension of Dijkstra's language (necessarily, since it is not monotonic with respect to $\sqsubseteq$; it is in fact the "+" operator of [31]). Nevertheless, the meaning that definitions 8, 9, and 10 give to the if construction **if** $([] i.\, B_i \rightarrow P_i)$ **fi** is exactly as before. We have

Theorem 6 *If P_i are feasible programs, then*

$$\text{if } ([\!]i.\ B_i \to P_i)\ \text{fi} <R> \ =\ (\vee i.\ B_i) \wedge (\wedge i.\ B_i \Rightarrow P_i <R>).$$

Proof: Let P be $([\!]i.\ B_i \to P_i)$. By definitions 9 and 10,

$$P<R>\ =\ (\wedge i.\ B_i \Rightarrow P_i <R>). \tag{6}$$

Hence because the P_i are feasible,

$$\neg P <\textit{false}>\ =\ \neg(\wedge i.\ B_i \Rightarrow \textit{false})\ =\ (\vee i.\ B_i). \tag{7}$$

The result now follows from (6), (7), and definition 10.
♡

Unfortunately, we must note in conclusion that because the construction **if** $\cdots$ **fi** is not monotonic, we have in general

$$P \sqsubseteq Q \qquad \textit{does not imply} \qquad \text{if } P \text{ fi} \sqsubseteq \text{if } Q \text{ fi}.$$

This limits its use in program development.

8 Positive applications of miracles

By definitions 7 and 6, miracles refine only to other miracles — and hence by Dijkstra's law never to programs. Thus if a specification *overall* is miraculous (we can check using theorem 4), the development is doomed.

In VDM, where specifications are written as predicate pairs like ours, the check for miracles is the "implementability test" [33, p. 134]. In Z [25], [49], [59], where specifications are single predicates corresponding to our implicit form of section 2.4, miracles cannot be written: definition 4 and theorem 4 show that single predicate specification statements are always feasible.

From a feasible beginning, miracles can arise through mistaken refinement tactics. As shown in section 6, the "improper division" of $x := 0$ by $x := y$ gives the miraculous x: $[\textit{true}\ ,\ y = 0]$. If we recognise the miracle then, we could stop there and try some other tactic; if we don't, we'll be stuck later. But the *rules* for such division (the weakest prespecification of [31]) are simpler now that soundness has been delegated to the unimplementable miracles: there is less need for "applicability conditions."

There is other potential for the deliberate use of miracles. Consider the following assignment, in which f is some function hard to calculate but easy to invert:

$$x := f(c) \tag{8}$$

And suppose in a variable y we might have the desired answer already. We can make the following refinements, in which both right hand sides are miracles:

$$x := f(c) \ \sqsubseteq \ c = f^{-1}(y) \rightarrow x := y \tag{9}$$
$$x := f(c) \ \sqsubseteq \ c \neq f^{-1}(y) \rightarrow x := f(c) \tag{10}$$

Neither (9) not (10) can be implemented on its own. Case (9) can be executed only when y does contain the desired answer already; case (10) can be executed only when it doesn't. But their $[\!]$ combination is *not* miraculous, and can always be executed:

$$(c = f^{-1}(y) \rightarrow x := y) \ [\!] \ (c \neq f^{-1}(y) \rightarrow x := f(c)) \tag{11}$$

Since $P \sqsubseteq Q$ and $P \sqsubseteq R$ implies $P \sqsubseteq Q[\!]R$ (easily shown from definitions 6 and 9), we have refined (8) to (11). Such developments are treated also in [2] and [54].

Another application is as follows. Ordinarily we limit the syntax of our concrete programming language so that miracles cannot be written in it: no specifications can appear, nor naked guarded commands. If we relax this restriction, allowing naked guarded commands, then operational reasoning suggests a *backtracking* implementation. For example, consider the following backtracking strategy for finding the position i of a value v in an array $a[0..N-1]$:

> Choose i at random from the range $0..N-1$, and evaluate $a[i] = v$. If equality holds, then terminate; otherwise, backtrack and try again.

We have this refinement:

$\quad\quad i\colon [a[i] = v]$

$\sqsubseteq \quad$ **if**
$\quad\quad\quad i := 0 \ [\!] \cdots [\!] \ i := N - 1;$
$\quad\quad\quad a[i] = v \rightarrow \textbf{skip}$
$\quad\quad$ **fi**

We are using the generalised **if** $\cdots$ **fi** of section 6, which here allows abortion if its body is miraculous; and the body is miraculous *only* when no branch of the alternation can avoid the miraculous behaviour to follow. In this context

if $\cdots$ **fi** resembles the "cut" of Prolog, allowing failure (preventing backtracking) if no solution is found within (beyond). If there is a successful branch, however, the implementation is obliged to find it: only then can it execute the second statement — which we could syntactically sugar, writing **force** $a[i] = v$. Note that the first statement can be written $i: [0 \leq i < N]$.

A third opportunity for exploiting miracles is in novel proof rules. We introduce for a moment the weaker relation $\leq$ between programs, which holds if for all predicates R

$$P < R > \wedge Q < true > \Rightarrow Q < R >$$

This is simply *partial* correctness. Now in the style of VDM we can consider a loop invariant to be a *statement*, rather than an assertion: any number of iterations of the loop body must refine the invariant statement I. The advantage is that we have easy reference to the initial state; our development law is

$$\begin{array}{lrcl} \text{If} & I & \leq & I;\ G \rightarrow S \\ \text{and} & X & \leq & I;\ \textbf{force}\ \neg G \\ \text{then} & X & \leq & I;\ \textbf{do}\ G \rightarrow S\ \textbf{od} \end{array}$$

We "explain" this rule as follows (but it is *proved* using weakest precondition semantics). The first condition requires preservation of the effect of I by one more execution of the body $G \rightarrow S$. If G holds, the body behaves like S; but if G fails (and therefore we should *not* execute S), the first condition still holds because $G \rightarrow S$ in that case is miraculous, refining anything (and **skip** in particular).

Similar reasoning applies to the second condition. For the result, we argue informally that

$$\begin{array}{rl} & X \\ \leq & I;\ \textbf{force}\ \neg G \\ \leq & \text{by induction over the first condition} \\ & I;\ G \rightarrow S;\ \cdots;\ G \rightarrow S;\ \textbf{force}\ \neg G \\ \leq & I;\ \textbf{do}\ G \rightarrow S\ \textbf{od} \end{array}$$

Take for example the following program, in which we calculate the sum s of an array a indexed by $0 \leq i < N$.

$$\begin{array}{rcl} X & = & s, n\colon [s = (\sum i : 0 \leq i < N : a[i])] \\ I & = & s, n\colon [s = (\sum i : 0 \leq i < n : a[i]) \wedge 0 \leq i \leq N] \\ G & = & n \neq N \\ S & = & s, n := s + a[n], n + 1 \end{array}$$

Because $I \sqsubseteq s, n := 0, 0$ (this is the initialisation), and because we can prove the conditions hold (using definitions and theorem 3), we have by our rule above

$$\begin{aligned} X &\leq I;\ \mathbf{do}\ n \neq N \to s, i := s + a[n], n + 1\ \mathbf{od} \\ &\leq s, n := 0, 0; \\ &\quad \mathbf{do}\ n \neq N \to s, i := s + a[n], n + 1\ \mathbf{od} \end{aligned}$$

9 Conclusion

We have extended Dijkstra's programming language with a construct allowing abstract programs, as predicate pairs, to be written within otherwise conventional "concrete" programs. The advantages are:

- Program development takes on the character of solving equations — well-established in mathematics generally. The transformation from abstract to concrete occurs within a single semantic framework.

- As lambda-expressions allow us to write functions without names (rather than the laboured "f **where** $f(x) = \cdots$") so we can write specifications directly, avoiding "P **where** $\cdots \Rightarrow P < \cdots >$." Instead of a lambda calculus, this leads to a refinement calculus.

- We gain *miracles* as an artefact of our extension, and there is increasing evidence that they simplify the development process. In [48] it is shown that applicability conditions for refinement can be simplified — or even removed altogether — because mistaken development steps simply lead to miracles from which eventually progress must cease. Also in [2], [31], [50], and more recently [54] it is argued that miracles simplify the theory. In [42] it is shown that miracles allow proof of certain data-refinements that were not provable previously.

- The lack of distinction between abstract and concrete programs allows their treatment as procedures to be made more uniform, in the sense of ALGOL-60: a procedure call, whether abstract or not, is equivalent to its text substituted in-line. This and the resulting treatment of parameters is explored in [43].

- The programmer's repertoire is increased by providing easy access to non-constructive idioms, for example: i: $[a[i] = v]$ finds the index i of value v in array a; m: $[l \leq m \leq h]$ chooses m between l and h.

- A ready connection is made with state-based specifications such as those of Z [25], [49], [59], allowing their systematic development into code.

A refinement calculus would be a collection of *laws*, each proved directly from weakest precondition definitions. They could be used, without further proof, in program developments — just as one uses a table of integrals in engineering. For example, one such law is

Assignment law: $\quad w\colon [post[w\backslash E][v_0\backslash v]\ ,\ post]$
$\sqsubseteq\quad w := E$

It is easily proved from definitions 3 and 6. A comprehensive collection of such laws is given and demonstrated in practice in [48].

Such a development style would be very close to VDM [33], where specifications are predicate pairs just as here. But Jones does not base VDM on the weakest precondition calculus, nor does he present a general refinement relation operating uniformly between all programs whether abstract or concrete (although he could do so). Another difference is our use of classical logic rather than the logic of partial functions [33], [12]. Jones does not treat miracles.

In the Z specification technique, specifications are given as single predicates corresponding to our "implicit preconditions". Thus where we write $n\colon [0 \leq n < n_0]$ for "decrease n, but not below 0," in Z one would write (omitting types)

$$\frac{n, n'}{0 \leq n' < n}$$

In Z there is no commitment to a *fixed* state (our $\vec{v}$); deliberately not, because this gives it the flexibility needed to build large specifications from their smaller components. Examples of large-scale Z specifications can be found in [25]. But when algorithmic structures are introduced — *i.e.*, once *development* begins — this lack of commitment becomes a hindrance.

Therefore one aim of our work is to provide a development method specifically for Z, by identifying the two specifications above then using the weakest precondition calculus to reach a concrete program. Another approach to Z development — derived from ours — is given in [34].

10 Acknowledgements

Back [5] first embedded specifications within programs using the weakest precondition calculus. His specifications — like those of Z — consist of one predicate only, and so he cannot take advantage of miracles. More recently Morris [50] presents independently the same extension of Back's work as we do; we have had useful discussions since discovering each other. Our refinement relation $\sqsubseteq$ is the same as theirs.

Meertens [39] also has developed these ideas, using predicate pairs, but gave them a different meaning: (in our notation) he defines

$$[pre, post] < R > \ \widehat{=} \quad \begin{array}{l} pre \Rightarrow (\exists \vec{v}.\, post) \\ \wedge \quad (\forall \vec{v}.\, post \Rightarrow R) \end{array}$$

But Meertens' definition does not have the property of lemma 2, which we consider to be fundamental; in general, for Meertens

$$[pre, post] < post > \ \neq \ pre.$$

Hehner [27] uses predicate pairs *for specifications* as we use specification statements, but he does not integrate the approach by giving them a weakest precondition semantics. He also uses the refinement ordering $\sqsubseteq$.

The earliest example of a formulation like ours for the weakest precondition of a specification seems to be Hoare's [29], where it is given as the axiomatic meaning of procedure calls. But he did not separate abstraction from procedure calling, as we have done (and discuss further in [43]). In [21, p. 153] also the definition can be found, again coupled to procedure calls.

The idea of using pre- and post-conditions to describe program behaviour is widespread, and its use in VDM is notable. In fact our approach is very close to VDM, and I hope identical in spirit. Jones does not however make his specifications "first-class citizens" as we do. An advantage of Jones's natural deduction style is perhaps its appeal to the wider audience of practising programmers, just as natural deduction in logic is so-called because it's more "natural." But we prefer the increased freedom of the axiomatic approach directly (in logic, too): it offers more scope to the experienced user, who can construct new laws (meta-theorems) to suit his taste and skill.

[31] provided the direct inspiration for treating specifications as programs; there similar results are obtained in the relational calculus. Miracles appear as partial relations, but are not discussed in detail.

Most recently, Nelson [54] has integrated specifications and programs, but his ordering over these objects differs from ours. In particular, it does not allow the reduction of non-determinism — an essential idea in program development. He discusses miracles at some length.

Much of this work was done in collaboration with Ken Robinson. I thank Rick Hehner, Joe Morris, Doaitse Swierstra, members of IFIP 2.1, and the referees for their very useful comments.

Specification Statements and Refinement

Carroll Morgan and Ken Robinson

Abstract

We discuss the development of executable programs from state-based specifications written in the language of first-order predicate calculus. Notable examples of such specifications are those written using the techniques Z and *VDM;* but our interest will be in the rigorous derivation of the algorithms from which they deliberately abstract. This is of course the role of a *development method.*

Here we propose a development method based on *specification statements* with which specifications are embedded in programs — standing in for developments "yet to be done." We show that specification statements allow description, development, and execution to be carried out within a *single* language: programs/specifications become hybrid constructions in which both predicates and directly executable operations can appear.

The use of a single language — embracing both high- and low-level constructs — has a very considerable influence on the development style, and it is that influence we will discuss: the specification statement is described, its associated calculus of refinement is given, and the use of that calculus is illustrated.

1 Introduction

In the Z [25, 49, 59] and VDM [33] specification techniques, descriptions of external behaviour are given by relating the "before" and "after" values of variables in a hypothetical program state. It is conventional to assume that the *external* aspects are treated by designating certain variables as containing initially the *input* values, and certain others as containing finally the *output* values. As development proceeds, structure is created in the program — and the specifications, at that stage more "abstract algorithms," come increasingly to refer to internal program variables as well. For example, we may at some stage wish to describe the operation of taking the square-root of some integer

---

Appeared in *IBM Jnl. Res. Dev. 31(5)* (Sept. 1987).
© Copyright 1987 by International Business Machines Corporation, reprinted with permission.

variable n; adopting the convention that n refers to the value of that variable *after* the operation, and n_0 to its value *before*, this description could be written:

$$n^2 = n_0 \tag{1}$$

Ordinarily, we would call the above a *specification*, because "conventional" computers do not execute (*i.e.*, find a valuation making true) arbitrary formulas of predicate logic (logic programming languages deal only with a restricted language of predicates).

Two notable features of our specification (1) above are its *non-determinism* and that it is *partial*. It is non-deterministic in the sense that for some initial values n_0 (*e.g.*, 4) there may be several appropriate final values n (± 2 in this case). It is partial in the sense that for some initial values (*e.g.*, 3) there are *no* appropriate final values. We will see below that our proposed development method makes this precise in the usual way (*e.g.*, of [17]): the non-determinism allows an implementation to return either result (either consistently or even varying from one execution to the next); and the implementor can *assume* that the initial value is a perfect square, providing a program whose behaviour is wholly arbitrary otherwise.

In presenting a development technique, we are not ignorant of the fact that VDM already has (or even *is*) one; rather we are concentrating our attention on Z, where development has been less well worked out. In this our aim is most definitely to propose a *light-weight* technique — as Z is itself — in which existing material is used as much as possible. Dijkstra's language [17] therefore was chosen as the target, because it has a mathematically attractive and above all simple semantic basis, and because it includes non-determinism naturally.

The *key* to a smooth development process — the subject of this paper — is we believe the integration of description and execution in one language. This is not achieved, as is so often proposed, by restricting our language to those specifications which are executable, and thus treating specifications as programs; instead we extend the language to allow ourselves to write programs which we cannot see at first how to execute: in effect we treat programs as specifications. It is precisely the lack of semantic distinction between the two that allows finally our smooth transition from abstract description to executable algorithm.

We will assume some familiarity with Dijkstra's weakest pre-condition concept and its associated guarded command programming language [17].

1.1 Weakest pre-conditions and specifications

In [17], Dijkstra introduces for program P and predicate R over the program variables, the *weakest pre-condition* of R with respect to P; he writes it

$$wp(P, R)$$

This weakest pre-condition is intended to describe exactly those states from which execution of P is guaranteed to establish R, and Dijkstra goes on to develop a small language by defining for its every construct precise syntactic rules for writing $wp(P, R)$ as a predicate itself. For example, the meaning of *assignment* in this language is defined as follows for variable x, expression E, and post-condition R:

$$wp(\text{``}x := E\text{''}, R) = R[x\backslash E]$$

The notation $[x\backslash E]$ here denotes syntactic replacement in R of x by E in the usual way (avoiding variable capture *etc.*). Thus

$$\begin{aligned}
& wp(\text{``}x := x-1\text{''},\ x \geq 0) \\
=\ & (x \geq 0)[x\backslash x-1] \\
=\ & (x-1) \geq 0 \\
=\ & x > 0
\end{aligned} \qquad (2)$$

We can *specify* a program P by giving *both* a pre-condition (not necessarily weakest) and a post-condition; our pre-condition and post-condition predicates we will usually call *pre* and *post*:

$$pre \Rightarrow wp(P, post) \qquad (3)$$

Informally, this is read "if *pre* is true, then execution of P must establish *post*"; formally, we regard the above as admitting only program texts P for which it is valid. Either way, it is a specification in the sense that it directs the implementor to develop a program with the required property.

Our point of divergence from the established style (3) is to write instead

$$[pre,\ post] \sqsubseteq P \qquad (4)$$

We take (3) and (4) as identical in meaning, but in (4) the constituents are exposed more clearly: $[pre,\ post]$ is the specification; $\sqsubseteq$ is the relation of refinement; and P is the program to be found. Thus we will read (4) as "the specification $[pre,\ post]$ is refined by P."

The principal advantage of the alternative style is that [pre , post] can take on a meaning independent of its particular use in (4) above: we will give it a weakest pre-condition semantics of its own. It is just this which removes the distinction between specification and program — not that they both are executable, but that they both are *predicate transformers*, being suitable first arguments to $wp(\ ,\ )$. Programs are just those specifications which we can execute directly.

The refinement relation $\sqsubseteq$ is likewise generalised, and we do this immediately below.

1.2 Refinement

In (4) we have introduced an explicit symbol "$\sqsubseteq$" for refinement, and we now give its precise definition (as given *e.g.*, in [27]):

Definition 1 *For programs P and Q, we say that P is refined by Q, written $P \sqsubseteq Q$, iff for all post-conditions* post:

$$wp(P,\ post) \Rightarrow wp(Q,\ post).$$

♡

We justify the above informally by noting that any occurrence of P in a (proved correct) program is justified by the truth of $wp(P, post)$ at that point, for some predicate *post*. No matter what *post* it is, the relation $P \sqsubseteq Q$ gives us $wp(Q, post)$ as well, so that Q is similarly justified: thus Q can replace P. Operationally, $P \sqsubseteq Q$ whenever Q resolves non-determinism in P, or terminates when P might not.

This refinement relation is independent of the notion of specification, and can be evaluated for *any* two constructs whose weakest pre-condition semantics are known. For example, we have in the guarded command language of [17]

$$\begin{array}{l} \text{if } a \leq b \rightarrow a := a - b \\ [\!]\ b \geq a \rightarrow b := b - a \\ \text{fi} \end{array}$$

$\sqsubseteq$ $\begin{array}{l} \text{if } a \leq b \rightarrow a := a - b \\ [\!]\ a \not\leq b \rightarrow b := b - a \\ \text{fi} \end{array}$

The first program is non-deterministic, executing either branch when $a = b$; the second program is a proper (*i.e.*, non-identical) refinement of it because this non-determinism has been removed. Such refinement relations between *programs* allow us to implement the non-deterministic program above in more conventional (deterministic) languages; we transcribe the deterministic refinement as follows:

```
IF a<=b THEN a:= a-b
        ELSE b:= b-a
END
```

1.3 Specification statements

From section 1.2 above, we can see that in formal terms we should have $[pre\ ,\ post] \sqsubseteq P$ iff for all R

$$wp([pre\ ,\ post]\,,\,R) \Longrightarrow wp(P,\,R) \tag{5}$$

But for this to have meaning, we must define its antecedent; as in the definition (2) above for assignment statements, we will express $wp([pre\ ,\ post]\,,\,R)$ as a syntactic transformation of the predicate R. We do this below, moving from simple to more general cases.

1.4 The simple case

In the simplest case we have two predicates *pre* and *post* each over the program variables in a single state. We have

Definition 2 *Let the vector of currently declared program variables be $\vec{v}$; for any predicates* pre, post, *and* R, *we define*

$$wp([pre, post],\,R) \quad = \quad pre \wedge (\forall\, \vec{v}\ .\ post \Rightarrow R)$$

♡

Note that our quantifiers always extend in scope to the first enclosing parentheses $(\forall . \cdots)$. As indicated, we will use $\vec{v}$ to refer to the vector of all program variables, and will not concern ourselves very much with how they are declared.

Section 2 discusses the consistency of definition 2 and formula (5); here we will justify the definition only informally. We regard $[pre, post]$ as a statement, and its first component *pre* describes the states in which its termination is guaranteed; thus *pre* is a necessary feature of our desired weakest pre-condition, and in fact appears as the first conjunct there. But the weakest pre-condition must guarantee more than termination: it must ensure that on termination, R holds. From the second component of $[pre, post]$, we know that $post$ describes the states in which it terminates — and so we require only that in all states described by $post$ the desired R holds as well: this is the second conjunct.

We now continue with some notational extensions and abbreviations.

1.4.1 Confining change

We allow a list of variables $\vec{w}$, in which appear all the variables which the statement can change; variables not in $\vec{w}$ must retain their initial values. The precise definition of $\vec{w} : [pre, post]$ is

Definition 3 *Let the vector of currently declared program variables be $\vec{v}$, and let $\vec{w}$ be a sub-vector of $\vec{v}$; for any predicates* pre, post, *and* R, *we define*

$$wp(\vec{w} : [pre, post], R) \;=\; pre \wedge (\forall \vec{w} \,.\, post \Rightarrow R)$$

♡

The only change from definition 2 is that the vector of quantified variables is now $\vec{w}$ rather than $\vec{v}$. Taking for example $\vec{v}$ to be "x, y", we have

$$\begin{aligned} & wp(x : [true, x = y], R) \\ =\; & true \wedge (\forall x \,.\, x = y \Rightarrow R) \\ =\; & R[x \backslash y]. \end{aligned}$$

Since also $wp(x := y, R) = R[x \backslash y]$, we have shown "$x : [true, x = y]$" and "$x := y$" to have the same meaning.

1.4.2 Initial values

So far, we can specify only that a certain relationship (*e.g.*, post) is to hold between the *final* values of variables. We now adjust our definition so that 0-subscripted variables in the second component of a specification statement can be taken as referring to the *initial* values of variables.

Definition 4 *Let the vector of currently declared program variables be $\vec{v}$, and let $\vec{w}$ be a sub-vector of $\vec{v}$; let* pre *and* R *as before be arbitrary predicates, and let* post *be a predicate referring optionally to 0-subscripted variables $\vec{v}_0$ as well. We define*

$$wp(\vec{w} : [pre, post], R) \;=\; pre \wedge (\forall \vec{w} \,.\, post \Rightarrow R)[\vec{v}_0 \backslash \vec{v}]$$

♡
— *provided* R *contains no 0-subscripted variables* $\vec{v}_0$.

By our definition we have *reserved* the use of 0-subscripts to denote initial values, and so must forego their use for other purposes: this is why R should contain no $\vec{v}_0$. It is possible, however, to take the view that in R also the variables $\vec{v}_0$ refer to initial values; this leads in fact to the weakest pre-*specification* of Hoare and He [31]. Josephs [34] has investigated this.

We note that if *post* does not refer to initial values, then definition 4 reduces to definition 3.

The substitution $[\vec{v}_0 \backslash \vec{v}]$ may require renaming of the bound variables $\vec{w}$, but this is often unnecessary; for example, taking $\vec{v}$ to be "x, y" as before, we have

$$\begin{aligned}
& wp(x : [true, x = x_0 + y_0], R) \\
= \; & true \land (\forall\, x\, .\, x = x_0 + y_0 \Rightarrow R)[x_0, y_0 \backslash x, y] \\
= \; & R[x \backslash x_0 + y_0][x_0, y_0 \backslash x, y] \\
= \; & R[x \backslash x + y].
\end{aligned}$$

This is of course $wp(x := x + y, R)$, as one would hope.

1.4.3 Implicit pre-conditions

If the pre-condition is omitted, we will supply a default condition for it as follows:

Definition 5 *Let the vector of currently declared program variables be* $\vec{v}$, *and let* $\vec{w}$ *be a sub-vector of* $\vec{v}$; *let* post *be a predicate referring optionally to 0-subscripted variables* $\vec{v}_0$. *We define*

$$\vec{w}: [post] \quad \text{abbreviates} \quad \vec{w}: [(\exists\, \vec{w}\, \bullet\, post)[\vec{v}_0 \backslash \vec{v}]\, ,\, post]$$

♡

Thus the *implicit* pre-condition is simply "it is possible to establish the post-condition". This is exactly the view taken in Z specifications generally, where only a single predicate is given; in our original square-root example (1) — writing it $n: [n^2 = n_0]$ — the implicit precondition is $(\exists\, n\, .\, n^2 = n_0)[n_0 \backslash n]$ which we can simplify to $(\exists\, k\, .\, k^2 = n)$. That is, termination is guaranteed only if n is a perfect square.

1.4.4 Generalised assignment

The assignment statement $x := E$ establishes the post-condition $x = E$ while changing only x — it has the same meaning, therefore, as the specification statement $x\colon [x = E[x\backslash x_0]]$ (in which the renaming $[x\backslash x_0]$ is necessary because occurrences of x in E are *initial* values). Exploiting this, we define below a *generalised* assignment statement in which the binary relation $=$ of ordinary assignment can be replaced by any binary relation desired.

Definition 6 *If "⊲" is a binary relation symbol, then for any variable x and expression E,*

$$x :\triangleleft E \quad \text{abbreviates} \quad x\colon [x \;\triangleleft\; E[x\backslash x_0]].$$

♡

Thus we have that

$$\begin{array}{ll} x :< x & \text{decreases } x; \text{ and that} \\ m :\in s & \text{chooses a member } m \text{ from the set } s. \end{array}$$

Note that in the second case, our implicit pre-condition is "the set s is not empty":

$$\begin{array}{rl} & m :\in s \\ = & m\colon [m \in s] \\ = & m\colon [(\exists m' \bullet m' \in s) \,,\, m \in s] \\ = & m\colon [s \neq \{\} \,,\, m \in s] \end{array}$$

This abbreviation was suggested by Jean-Raymond Abrial.

1.4.5 Invariants

Often a formula appears as a conjunct in both the pre- and the post-conditions, thus making it an *invariant* of the statement. The following convention, suggested in [27], allows us to write it only once; we abbreviate $[pre \wedge I \,,\, I \wedge post]$ by

$$[pre \,,\, I \,,\, post]$$

Thus $[pre \,,\, I \,,\, post] \sqsubseteq Q$ iff

$$pre \wedge I \implies wp(Q, I \wedge post).$$

The above convention is useful when developing loops, as we will see in section 3.

2 The refinement theorems

The following theorems justify our choice of semantics for the specification statement. (Their full proofs may be found in [44].) The first theorem shows that for every specification there is a specification statement that satisfies it trivially.

Theorem 1 *If $\vec{u}$ and $\vec{w}$ partition the vector $\vec{v}$ of program variables, then for any predicates* pre *and* post

$$pre \wedge \vec{v} = \vec{v}_0 \quad \Longrightarrow \quad wp(\vec{w} : [pre, post],\ post \wedge \vec{u} = \vec{u}_0)$$

Proof *(outline):* *The result follows by straightforward application of definition 4 and predicate calculus, except for the possible occurrences of 0-subscripted variables in* $post \wedge \vec{u} = \vec{u}_0$. *Since these are not program variables (we never declare e.g. x_0 in a program), we can avoid the problem by a systematic renaming, proving instead that*

$$pre \wedge \vec{v} = \vec{v}_1 \quad \Longrightarrow \quad wp(\vec{w} : [pre, post],\ post[\vec{v}_0 \backslash \vec{v}_1] \wedge \vec{u} = \vec{u}_1)$$

This technique is used also in the proof of theorem 3 in section 5.1, given in full.
♡

The consistency mentioned in section 1.4 follows easily from the above, taking $\vec{w} = \vec{v}$ and *post* free of $\vec{v}_0$; clearly other specialisations are profitable as well.

The complementary problem is refining further a given specification statement; the following theorem shows how this can be done.

Theorem 2 *If $\vec{w}$ and $\vec{u}$ partition the program variables $\vec{v}$, and if*

$$pre \wedge \vec{v} = \vec{v}_0 \quad \Longrightarrow \quad wp(P,\ post \wedge \vec{u} = \vec{u}_0)$$

then

$$\vec{w} : [pre, post] \sqsubseteq P$$

Proof *(outline):* *The proof again simply applies definitions, this time definitions 1 and 4; the 0-subscripts are avoided as before.*
♡

To summarise: theorem 1 shows that $\vec{w}: [pre\,,\,post]$ is always *a* solution to the specification (of P):

$$pre \wedge \vec{v} = \vec{v}_0 \quad \Longrightarrow \quad wp(P,\, post \wedge \vec{u} = \vec{u}_0)$$

Theorem 2 shows it to be *more general* than any other solution; thus overall we have that is it the most general solution.

3 The refinement calculus

We now move to our main concern. With the definitions of section 1 we can mix specifications and executable constructs freely, and program development becomes a process of transformation within the one framework. But this is only the beginning — the definitions supply the "first principles" from which more specialised techniques spring, and we can use these derived *laws of refinement* directly in our development of programs. Each law is designed to introduce a particular feature into our final program, and the process overall comes to resemble the *natural deduction* style of formal proof, where our goals are not axioms but rather directly executable constructs (the Vienna Development Method [33] has a similar flavour).

We will present the laws in the form

$$\frac{before\text{-}refinement}{after\text{-}refinement} \quad side\text{-}condition$$

and by this we mean: "if *side-condition* is universally valid, then

$$before\text{-}refinement \sqsubseteq after\text{-}refinement"$$

Often, there is no side-condition — this indicates that the stated refinement always obtains.

3.1 Strengthening the specification

Generally speaking, refinement *strengthens* a specification, and it is characteristic of our refinement calculus that no check is made against strengthening a specification too much (a notable difference from VDM). The advantage of this is simplicity of the laws (law 11 provides a striking example); a disadvantage is that unproductive refinement steps may go longer unnoticed. But there is no danger of invalidity resulting from over-strengthened specifications, for we will see that they can never provably be refined to executable code.

There is a simple *feasibility test* that can be applied to any specification, and its failure predicts the failure of the refinement process: we simply check that the specification satisfies Dijkstra's *Law of the Excluded Miracle* [17, p. 18] (paraphrased)

"For all executable programs P,

$$wp(P, false) = false"$$

If the specification failed this law, then so would any refinement of it; and since no *executable* program fails the law, we are forced to conclude that such a specification can never be refined to an executable program. For specifications, direct calculation yields that $\vec{w}: [pre , post]$ is feasible iff $pre \Rightarrow (\exists \vec{w} . post)[\vec{v}\backslash\vec{v}_0]$. This was first pointed out by Robinson [58].

The essence of our advantage is therefore that our laws do *not* force us implicitly to apply a feasibility test at their every application: very often the correctness of a development step is obvious. Further discussion on this topic can be found in [44].

Our first two laws deal with weakening the pre-condition and/or strengthening the postcondition of a specification.

Law 1 *Weakening the pre-condition; the new specification is more robust than the old (*i.e., *it terminates more often):*

$$\frac{\vec{w}: [pre , post]}{\vec{w}: [pre' , post]} \quad pre \Rightarrow pre'$$

♡

For example, $n: [n > 0 , n = n_0 - 1] \sqsubseteq n: [n \geq 0 , n = n_0 - 1]$.

Law 2 *Strengthening the post-condition; the new specification allows less choice than the old:*

$$\frac{\vec{w}: [pre , post]}{\vec{w}: [pre , post']} \quad pre \Rightarrow (\forall \vec{w} \bullet post' \Rightarrow post)[\vec{v}_0\backslash\vec{v}]$$

♡

For example, $n: [true , n \geq 0] \sqsubseteq n: [true , n > 0]$.

It is worth noting that a special case of law 2 occurs when $\vec{v}$ and $\vec{w}$ are the same; then we have for the side-condition

$$pre \implies (\forall \vec{v} \bullet post' \implies post)[\vec{v}_0\backslash\vec{v}]$$

Renaming $\vec{v}$ to $\vec{v}_0$ throughout, this is equivalent to

$$pre[\vec{v}\backslash\vec{v}_0] \implies (\forall \vec{v} \bullet post' \implies post)[\vec{v}_0\backslash\vec{v}][\vec{v}\backslash\vec{v}_0]$$

which we may simplify to

$$pre[\vec{v}\backslash\vec{v}_0] \implies (\forall \vec{v} \bullet post' \implies post)$$

The quantifier $\forall \vec{v}$ can be discarded since the antecedent contains no $\vec{v}$, and propositional calculus then gives us as our special case the appealing

$$pre[\vec{v}\backslash\vec{v}_0] \wedge post' \implies post$$

Law 3 *Restricting change; the new specification can change fewer variables than the old:*

$$\frac{\vec{w}, x: [pre\ ,\ post]}{\vec{w}: [pre\ ,\ post]}$$

♡

For example, $x, y: [x = y_0] \sqsubseteq x: [x = y_0]$.

In law 4 below, we use the compact symbols |[and]|, instead of the more conventional **begin** and **end**, to delimit the scope of local variable declarations.

Law 4 *Introducing fresh local variables (where "fresh" means not otherwise occurring free):*

$$\frac{\vec{w}: [pre\ ,\ post]}{|[\textbf{var } x;\ \ \vec{w}, x: [pre\ ,\ post]\ ]|} \quad x \text{ is a fresh variable}$$

♡

For example, $f: [f = n!] \quad \sqsubseteq \quad |[\textbf{var } i;\ \ f, i: [f = n!]\ ]|$.

3.2 Introducing executable constructs

The following laws allow us to introduce constructs from our target programming language.

Law 5 *Introducing* **abort**:

$$\frac{\vec{w}\colon [\mathit{false}\ ,\ \mathit{post}]}{\mathbf{abort}}$$

♡

Since **abort** $\sqsubseteq P$ for any P, we can by transitivity of $\sqsubseteq$ have *any* program as the target of law 5. Thus for any predicate $\mathit{difficult}(n)$, we still have the easy refinement $n\colon [n < 0 \wedge n > 0\ ,\ \mathit{difficult}(n)] \sqsubseteq n := 17$.

Law 6 *Introducing* **skip**:

$$\frac{\vec{w}\colon [\mathit{post}[\vec{v}_0\backslash\vec{v}]\ ,\ \mathit{post}]}{\mathbf{skip}}$$

♡

For example, $x, y\colon [x = y\ ,\ x = y_0] \sqsubseteq \mathbf{skip}$.

Law 7 *Introducing assignment:*

$$\frac{\vec{w}\colon \left[\mathit{post}[\vec{v}_0, \vec{w}\backslash\vec{v}, \vec{E}]\ ,\ \mathit{post}\right]}{\vec{w} := \vec{E}}$$

♡

For example, $x\colon [\mathit{true}\ ,\ x = x_0 + y] \sqsubseteq x := x + y$.

The next two laws are the predicate transformer equivalents of the *weakest pre-specification* and *weakest post-specification* constructions of Hoare and He [31], with which one can "divide" one specification A by another B, leaving a specification Q such that

$$
\begin{array}{lll}
A \sqsubseteq Q;\ B & \text{(law 8: weakest pre-specification)} \\
A \sqsubseteq B;\ Q & \text{(law 9: weakest post-specification)}
\end{array}
$$

Law 8 *Introducing sequential composition (weakest pre-specification):*

$$\frac{\vec{w}\colon [pre\ ,\ post]}{\vec{w}\colon [pre\ ,\ wp(P,\ post)]\ ;\ P} \qquad \vec{w}\colon [true] \sqsubseteq P$$

♡

The side condition $w\colon [true] \sqsubseteq P$ can be read "P changes only $\vec{w}$". For example, we have

$$x, y\colon [true\ ,\ x = y + 1]$$

$$\sqsubseteq\ \ x, y\colon [true\ ,\ x = 2]\ ;$$
$$y := 1$$

Law 9 *Introducing sequential composition (weakest post-specification):*

$$\frac{\vec{w}\colon [pre\ ,\ post]}{\begin{array}{l}\vec{w}\colon [pre\ ,\ mid]\ ;\\ \vec{w}\colon [mid\ ,\ post]\end{array}} \qquad mid,\ post\ \text{contain no free}\ \vec{v}_0$$

♡

For example,

$$x\colon [true\ ,\ x = y + 1]$$

$$\sqsubseteq\ \ x\colon [true\ ,\ x = y]\ ;$$
$$x\colon [x = y\ ,\ x = y + 1]$$

Law 9 can be generalised to the case in which variables $\vec{v}_0$ do appear (as shown in [40]); in that case, one has effectively supplied in mid the first component of the sequential composition. For our larger example to follow (section 4), we need only the simpler version.

In laws 10 and 11, we use a quantifier-like notation for generalised disjunction and alternation: if I for example were the set $\{1..n\}$, then $(\bigvee i\colon I\ .G_i)$ would abbreviate $G_1 \vee \cdots \vee G_n$, and **if** $(\;[\!]\ i\ .G_i \rightarrow S_i)$ **fi** would abbreviate

if $G_1 \rightarrow S_1$
$[\!]\ \ \ldots$
$[\!]\ \ G_n \rightarrow S_n$
fi

Law 10 *Introducing alternation* (**if**):

$$\frac{\vec{w}\colon [pre \wedge (\bigvee i\colon I \,.\, G_i) \;,\; post]}{\textbf{if} \;(\![]\; i\colon I \,.\, G_i \to \vec{w}\colon [pre \wedge G_i \;,\; post])\; \textbf{fi}}$$

♡

The predicate *pre* is that part of the pre-condition irrelevant to the case distinction being made by the guards G_i: it is passed on to the branches of the alternation. For example, taking *pre* to be true, we have

$$y\colon [x = 0 \vee x = 1 \;,\; x + y = 1]$$
$\sqsubseteq \quad$ **if** $x = 0 \;\to\; y\colon [x = 0 \;,\; x + y = 1]$
$\quad\;\;[]\;\; x = 1 \;\to\; y\colon [x = 1 \;,\; x + y = 1]$
fi

$\sqsubseteq \quad$ **if** $x = 0 \;\to\; y := 1$
$\quad\;\;[]\;\; x = 1 \;\to\; y := 0$
fi

Law 11 *Introducing iteration* (**do**):

$$\frac{\vec{w}\colon [true \;,\; inv \;,\; \neg(\bigvee i\colon I \,.\, G_i)]}{\textbf{do} \;(\![]\; i\colon I \,.\, G_i \to \vec{w}\colon [G_i \;,\; inv \;,\; 0 \leq var < var_0])\; \textbf{od}}$$

♡

The predicate *inv* is of course the loop invariant, and the expression *var* is the variant. We use var_0 to abbreviate $var[\vec{v}\backslash\vec{v}_0]$.

An example of law 11 is given in section 4; for now, we note that *inv* can be any predicate and *var* any integer-valued expression. Surprisingly, there are no side-conditions — a bad choice of *inv* or *var* or indeed G_i simply results in a loop body from which no executable program can be developed (see the remarks in section 3.1).

Law 11 is proved in section 5.

4 An example: square root

For an example, we take the square-root development of [17, pp. 61–65]; but our development here will be deliberately terse, because we are illustrating not how to *find* such developments (properly the subject of a whole book), but rather how experienced programmers could *record* such a development.

4.1 Specification

We are given a non-negative integer sq; we must set the integer variable rt to the greatest integer not exceeding $\sqrt{sq}$, where the function $\sqrt{}$ takes the non-negative square root of its argument.

4.2 Specification

$rt := \lfloor \sqrt{sq} \rfloor$

$\lfloor x \rfloor$ — the "floor" of x — is the greatest integer not exceeding x.

4.3 Refinement

We assume of course that $\sqrt{}$ is unavailable to us, and proceed as follows to eliminate it from our specification; we eliminate $\lfloor\ \rfloor$ also. "Stacked" predicates denote conjunction.

$$
\begin{array}{rl}
& rt := \lfloor \sqrt{sq} \rfloor \\
= & rt\colon [\, rt = \lfloor \sqrt{sq} \rfloor \,] \qquad\qquad\qquad \text{definition 6} \\
= & rt\colon [\, sq \geq 0 \,,\ rt = \lfloor \sqrt{sq} \rfloor \,] \qquad\qquad \text{definition 5} \\
= & rt\colon [\, sq \geq 0 \,,\ rt \leq \sqrt{sq} < rt+1 \,] \qquad \text{definition of } \lfloor\ \rfloor \\
\sqsubseteq & rt\colon \left[\, sq \geq 0 \,,\ \begin{array}{c} 0 \leq rt \\ rt^2 \leq sq < (rt+1)^2 \end{array} \,\right] \qquad \text{law 2}
\end{array}
$$

4.4 Refinement

Using laws 4 and 2, we introduce a new variable ru, and strengthen the post-condition; our technique will be to approach the result from above (ru) and below (rt):

$$\sqsubseteq \quad |[\textbf{var } ru \; . \\
\quad\quad rt, ru: \left[sq \geq 0 \; , \; \begin{array}{c} 0 \leq rt \\ rt^2 \leq sq < ru^2 \\ rt + 1 = ru \end{array} \right] \\
]|$$

We now work on the inner part.

4.5 Refinement

Anticipating use of $rt+1 \neq ru$ as a loop guard we concentrate on the remainder of the post-condition, using law 9 with $mid = \left(\begin{array}{c} 0 \leq rt < ru \\ rt^2 \leq sq < ru^2 \end{array} \right)$ to proceed:

$$\sqsubseteq \quad rt, ru: \left[sq \geq 0 \; , \; \begin{array}{c} 0 \leq rt < ru \\ rt^2 \leq sq < ru^2 \end{array} \right] ; \tag{6}$$

$$rt, ru: \left[\begin{array}{cc} 0 \leq rt < ru & 0 \leq rt < ru \\ rt^2 \leq sq < ru^2 \; , & rt^2 \leq sq < ru^2 \\ & rt + 1 = ru \end{array} \right]$$

Using laws 1 and 7, we can show that for the first component of the sequential composition above — establishing mid, to become the loop invariant — we have

$$\sqsubseteq \; rt, ru := \; 0, sq + 1$$

We now concentrate on the second component.

4.6 Refinement

We now introduce the loop, rewriting the second component of the sequential composition (6) to bring it into the form required by law 11; writing inv now for our mid above, we have

$$= \quad rt, ru: [true \; , \; inv \; , \; rt + 1 = ru]$$

and then by 11, with variant $ru - rt$, we proceed

$\sqsubseteq$ **do** $rt + 1 \neq ru \rightarrow$
         rt, ru: $[rt + 1 \neq ru\ ,\ inv\ ,\ 0 \leq ru - rt < ru_0 - rt_0]$
    od

4.7 Refinement

For the loop body, we use law 4 again to introduce a local variable rm to "chop" the interval $rt..ru$ in which the result lies:

$\sqsubseteq$ $|[$ **var** rm;
       rt, ru, rm: $[rt + 1 \neq ru\ ,\ inv\ ,\ 0 \leq ru - rt < ru_0 - rt_0]$
   $]|$

We first choose rm between rt and ru, using law 9 then law 3 twice to develop:

$\sqsubseteq$ rm: $[rt + 1 \neq ru\ ,\ inv\ ,\ rt < rm < ru]$;
     rt, ru: $[rt < rm < ru\ ,\ inv\ ,\ 0 \leq ru - rt < ru_0 - rt_0]$

Then with laws 1 and 7, we quickly dispose of the first component, deciding to make our choice of rm divide the interval evenly:

$\sqsubseteq rm := (rt + ru) \div 2$

We proceed with the second component.

4.8 Refinement

The natural case analysis is now to consider $rm^2 \leq sq$ versus $rm^2 > sq$; accordingly, with law 10, we so divide our task and immediately apply law 3 to each case; we have

$\sqsubseteq$ **if** $rm^2 \leq sq \rightarrow rt$: $\left[\begin{array}{c} rt < rm < ru \\ rm^2 \leq sq \end{array}\ ,\ inv\ ,\ 0 \leq ru - rt < ru_0 - rt_0 \right]$

     $[]\ \ rm^2 > sq \rightarrow rt$: $\left[\begin{array}{c} rt < rm < ru \\ rm^2 > sq \end{array}\ ,\ inv\ ,\ 0 \leq ru - rt < ru_0 - rt_0 \right]$
     fi

For the first branch, we have by law 7

$\sqsubseteq rt := rm$

For the second branch, we have similarly

$\sqsubseteq ru := rm$

This completes our development.

4.9 Consolidation: the implementation

Developments in this style generate a tree structure in which children collectively refine their parents; to obtain the program "neat," we simply flatten the tree. For the square root program, the result is as follows:

$$
\begin{array}{l}
|[\ \textbf{var}\ ru; \\
\quad rt, ru := 0, sq + 1; \\
\quad \textbf{do}\ rt + 1 \neq ru\ \rightarrow \\
\qquad |[\ \textbf{var}\ rm; \\
\qquad\quad rm := (rt + ru) \div 2; \\
\qquad\quad \textbf{if}\ rm^2 \leq sq\ \rightarrow\ rt := rm \\
\qquad\quad []\ \ rm^2 > sq\ \rightarrow\ ru := rm \\
\qquad\quad \textbf{fi} \\
\qquad]| \\
\quad \textbf{od} \\
]|
\end{array}
$$

It is to be stressed that this consolidated presentation is not to be carried off as the only relic of our development. The development itself must remain as a record of design steps taken and their justifications (and in industrial practice, of who took them!). Mistakes will still be made, and corrections applied; only when a complete record is kept can we make those corrections reliably, without introducing further errors — and learn from the process.

5 Derivation of laws

In this section we will prove the laws 2 and 11 of section 3. We do this for several reasons: to reassure the reader, who may doubt their validity; to demonstrate the use of the weakest pre-condition formula for specifications; and to suggest that the collection of laws can easily be extended by similar proofs.

5.1 Proof of law 2

Law 2 allows us to strengthen the post-condition of a specification; in simplest terms, this means replacing *post* by *post'* as long as we know that $post' \Rightarrow post$. The side-condition is weaker than this, however: it takes both the pre-condition and changing variables into account, making the law more widely applicable.

In the proof below, we will assume that free-standing formulae are *closed* — that is, that their free variables are implicitly quantified (universally). It is this that will allow us to rename variables when necessary.

Theorem 3 *Proof of law 2: if the following side-condition holds*

$$pre \Longrightarrow (\forall \vec{w} \bullet post' \Rightarrow post)[\vec{v}_0 \backslash \vec{v}]$$

then so does this refinement:

$$\vec{w}: [pre\ ,\ post] \sqsubseteq \vec{w}: [pre\ ,\ post']$$

Proof *By theorem 2, we need only show*

$$pre \wedge \vec{v} = \vec{v}_0 \Longrightarrow wp(\vec{w}: [pre\ ,\ post'], post \wedge \vec{u} = \vec{u}_0)$$

Since in definition 4 the predicate R must not contain $\vec{v}_0$, we rename those above to $\vec{v}_1$ (we may do this because the formula is closed); we must show

$$pre \wedge \vec{v} = \vec{v}_1 \Longrightarrow wp(\vec{w}: [pre\ ,\ post'], post \wedge \vec{u} = \vec{u}_1)$$

Definition 4 is now applied; we must show

$$pre \wedge \vec{v} = \vec{v}_1 \Longrightarrow pre \wedge (\forall \vec{w} \bullet post' \Rightarrow post \wedge \vec{u} = \vec{u}_1)[\vec{v}_0 \backslash \vec{v}]$$

Clearly we can remove the conjunct pre in the consequent, because it occurs in the antecedent; we can remove $\vec{u} = \vec{u}_1$ because $\vec{u}$ and the quantified $\vec{w}$ are disjoint, and $\vec{v} = \vec{v}_1$ appears in the antecedent. It remains to prove

$$pre \wedge \vec{v} = \vec{v}_1 \Longrightarrow (\forall \vec{w} \bullet post' \Rightarrow post)[\vec{v}_0 \backslash \vec{v}]$$

And this follows directly from the side-condition.

♡

5.2 Proof of law 11

We will deal with the following restricted version of law 11, in which we consider a single guard only and take $\vec{v}$ and $\vec{w}$ the same; we must show

$$\frac{[true \, , \, inv \, , \, \neg guard]}{\begin{array}{l} \text{do} \\ \quad guard \; \rightarrow \; [guard \, , \, inv \, , \, 0 \leq var < var_0] \\ \text{od} \end{array}}$$

Our proof is based on the loop semantics given in [17]; we will show that for $k \geq 1$

$$inv \wedge (guard \Rightarrow var < k) \Longrightarrow H_k(inv \wedge \neg guard) \tag{7}$$

From this will follow

$$\begin{array}{rl} & inv \\ = & inv \wedge (guard \Rightarrow (\exists\, k \geq 1 \; \bullet \; var < k)) \\ = & (\exists\, k \geq 1 \; \bullet \; inv \wedge (guard \Rightarrow var < k)) \\ \Rightarrow & (\exists\, k \; \bullet \; H_k(inv \wedge \neg guard)) \\ = & wp(\text{do}\cdots\text{od}, inv \wedge \neg guard) \end{array}$$

Thus by theorem 1 we will have as required that

$$[inv \, , \, inv \wedge \neg guard] \sqsubseteq \text{do}\cdots\text{od}$$

It remains therefore to prove (7), and this we will do by induction over k. We note first that $H_0 = inv \wedge \neg guard$, and continue by direct calculation (writing pre' for $pre[\vec{v}\backslash\vec{v}']$ etc., and H_k for $H_k(inv \wedge \neg guard)$):

$$\begin{array}{rl} & H_1 \\ = & H_0 \vee \left(\begin{array}{c} guard \\ guard \wedge inv \\ (\forall\, \vec{v} \; \bullet \; inv \wedge 0 \leq var < var_0 \Rightarrow H_0)[\vec{v}_0\backslash\vec{v}] \end{array} \right) \\ = & H_0 \vee \left(\begin{array}{c} guard \wedge inv \\ (\forall\, \vec{v}' \; \bullet \; inv' \wedge 0 \leq var' < var \Rightarrow H_0') \end{array} \right) \\ \Leftarrow & (\neg guard \wedge inv) \vee (guard \wedge inv \wedge var < 1) \\ = & inv \wedge (guard \Rightarrow var < 1) \end{array}$$

43

Our inductive step now concludes the argument:

H_{k+1}

$= H_0 \lor \left(\begin{array}{c} guard \land inv \\ (\forall \vec{v}' \bullet inv' \land 0 \leq var' < var \Rightarrow H_k') \end{array} \right)$

$\Leftarrow H_0 \lor \left(\left(\forall \vec{v}' \bullet \begin{array}{c} guard \land inv \\ inv' \land 0 \leq var' < var \end{array} \Rightarrow \left(\begin{array}{c} inv' \\ guard' \Rightarrow var' < k \end{array} \right) \right) \right)$

$\Leftarrow H_0 \lor \left(\begin{array}{c} guard \land inv \\ var < (k+1) \end{array} \right)$

$= inv \land (guard \Rightarrow var < (k+1))$

♡

The puzzling thing about law 11 is that it has no side-condition, whereas one might expect to find the condition

$guard \land inv \Rightarrow 0 \leq var$

But closer inspection reveals that whenever the above formula fails, the loop body is infeasible: it must terminate (since $guard \land inv$ holds initially) and must establish $0 \leq var < 0$ (since $0 \not\leq var$ holds initially). By the law of the excluded miracle (see [17]), no executable program can do this — the refinement, though valid, is barren.

For the practising developer, perhaps the side-condition should be explicit; indeed, law 11 can be rewritten this way, with the $0 \leq var$ dropped from the post-condition of the loop body. For the historical record of our development however, we want to prove the very minimum necessary — and feasibility is of no interest. There would be no program, and hence no record, if a feasibility check would have failed.

6 Conclusion

We have claimed that the integration of specifications and executable programs improves the development process. In earlier work [44], the point was made that all the established techniques of refinement are of course still applicable; their being based on weakest pre-condition semantics automatically makes them suitable for *any* construct so given meaning. Indeed an immediate but modest application of this work is our writing for example "choose e from s" directly in our development language as "$e :\in s$".

The refinement calculus is a step further. We are not claiming that it makes algorithms easier to discover, although we hope that this will be so; but it clearly does make it easier to avoid trivial mistakes in development and to keep a record of the steps taken there. A professional approach to software development must record the development *process,* and it must do so with mathematical rigor. We propose the refinement calculus for that at least.

Another immediate possibility is the systematic treatment of Z "case studies" as exercises in development, and we hope to learn from this. (There are a large number of case studies collected in [48].) Such systematic development is already underway for example at the IBM Laboratories at Hursley Park, UK [55].

The techniques of *data* refinement, in which high-level data structures (sets, bags, functions ...) are replaced with structures of the programming language (arrays, trees ...), fit extremely well into this approach. Also facilitated is the introduction of procedures and functions into a development: the body of the procedure is simply a specification statement "yet to be refined," and the meaning of procedures can once more be given by the elegant *copy rule* of Algol-60. These ideas are explored in [40] and [58], and we hope to publish them more widely.

7 Acknowledgements

It is clear our approach owes its direction to the steady pressure exerted by the work of Abrial, Back, Dijkstra, Hoare, and Jones. More direct inspiration came from the *weakest pre-specification* work of Hoare and He [31], who provide a relational model and a calculus for development; they strongly advocate the *calculation* of refinements as an alternative to refinements proposed then proved. Robinson [58] has done earlier work on the refinement calculus specifically.

We believe the earliest embedding of specifications within Dijkstra's language of weakest pre-conditions to be that reported in Back's thesis [5], and to him we freely give the credit for it. His *descriptions* are single predicates, rather that the predicate pairs we use here, and he gives a very clear and comprehensive presentation of the resulting refinement calculus. Our work extends his in that we consider predicate *pairs,* as does VDM, but — unlike VDM — we do not require those pairs always to describe feasible specifications. Because of this, we obtain a significant simplification in the laws of our refinement calculus.

In [39] L. Meertens explores similar ideas, and we are grateful to him for making us aware of Back's work.

We have benefited from collaboration with the IBM Laboratory at Hursley Park; the joint project [55] aims to transfer research results directly from university to development teams in industry.

Morris [50] independently has taken a similar approach to ours (even to allowing infeasible *prescriptions*); we recommend his more abstract view, which complements our own.

To the referees, and to Stephen Powell of IBM, we are grateful for their helpful suggestions.

Procedures, Parameters, and Abstraction: Separate Concerns

Carroll Morgan

Abstract

The notions of *procedures*, *parameters*, and *abstraction* are by convention treated together in methods of imperative program development. Rules for preserving correctness in such developments can be complex.

We show that the three concerns can be separated, and we give simple rules for each. Crucial to this is the ability to embed *specification* — representing abstraction — directly within programs; with this we can use the elegant *copy rule* of ALGOL-60 to treat procedure calls, whether abstract or not.

Our contribution is in simplifying the use of the three features, whether separately or together, and in the proper location of any difficulties that do arise. The *aliasing* problem, for example, is identified as a "loss of monotonicity" with respect to program refinement.

1 Introduction

In developing imperative programs one identifies points of procedural abstraction, where the overall task can be split into subtasks each the subject of its own development subsequently. Integration of the subtasks is accomplished ultimately by parametrized procedure calls in the target programming language. We argue here that these concerns — *procedures, parametrization*, and *abstraction* — can be separated, and that the result is of practical utility.

Abstraction identifies a coherent algorithmic activity that can be split from the main development process; conventionally, a procedure call is left at the point of abstraction, and its necessary properties become the specification of the procedure body. Instead, we leave the specification itself at the point of abstraction, with no *a priori* commitment to a procedure call.

Appeared in *Sci. Comp. Prog. 11* (1988).
© Copyright 1988, Elsevier Science Publishers B.V. (North Holland), reprinted with permission.

Procedure call we treat as simple substitution of text for a name, not caring whether we substitute programming language code (as we would in the final program) or a specification (as we would in a high-level design).

Parametrization we treat as a substitution mechanism that can be applied uniformly to specifications or to program language code, whether or not a procedure call occurs there.

The aim is to give a simple *orthogonal* set of rules for treating each concern. Existing practice is in most cases easily realised by appropriate combinations of the rules; but the independence allows greater freedom than before.

2 Procedure call

We return to the simple view, taken in the ALGOL-60 (revised) report [53], that procedure calls are to be understood via a *copy rule:* a program that calls a procedure is equivalent to one in which the procedure name is replaced by the text of the procedure body. In the examples to follow, we declare (parameterless) procedures using

procedure *name* $\widehat{=}$ *body*

With the copy rule, therefore, we have the equality indicated in the following example:

$$
\boxed{\begin{array}{l}\textbf{procedure } \textit{Inc} \ \widehat{=} \ \ x := x + 1 \\ \ldots \\ x := 0; \\ \textit{Inc}; \\ \textbf{write } x\end{array}} \quad = \quad \boxed{\begin{array}{l}x := 0; \\ x := x + 1; \\ \textbf{write } x\end{array}} \qquad (1)
$$

The technique has impeccable credentials; it is for example strongly (and deliberately) related to the following one-point rule of predicate calculus:

$$(\forall x \bullet x = T \Rightarrow P) \quad \Rightarrow \quad P[x\backslash T]$$

We write quantifications within parentheses () which delimit their scope, and use a spot • to separate the binding variable from the body. In the formula above, T is some term not containing x free, P a predicate, and $P[x\backslash T]$ the result of replacing x by T in P. We assume that the substitution $[x\backslash T]$ is defined so that it avoids variable capture; similar care is needed with the copy rule.

But the copy rule gives the meaning only of programs written entirely in a programming language. In contrast, the modern "step-wise" approach to program development introduces hybrid programs in which names denote program fragments "yet to be implemented". One understands the effect of these fragments in terms of their specification — abstracting from the detail of implementation — using rules specifically for procedure call such as those given in [29], [21], and [22]. The simple copy rule cannot be applied, for there is not yet program text to copy.

In [29], for example, one finds a *Rule of adaptation*, in the style of the rules of [28], with which procedures specified by pre- and post-conditions can be proved to have been used correctly in a calling program. There is also given in [29] a *Rule of substitution* for dealing separately with the effects of parameter passing. In the more recent [22] and [21], combined rules treat procedures — as specifications — with their parameters, all at once.

Here we reverse the trend, not only retaining the earlier view [29], which separates procedure calling (*adaptation*) from parameter passing (*substitution*), but also splitting procedure call from procedural abstraction. For procedure calls, therefore, we retain only the copy rule of ALGOL 60 [53, 4.7.3.3]:

> ... the procedure body ... is inserted in place of the procedure statement ... If the procedure is called from a place outside the scope of any non-local quantity of the procedure body, the conflicts between the identifiers inserted through this process of body replacement and the identifiers whose declarations are valid at the place of the procedure statement ... will be avoided through suitable systematic changes of the latter identifiers.

3 Procedural abstraction

We take the axiomatic view: a procedural abstraction is described by a predicate pair comprising a *pre-condition* and a *post-condition*, both built (mainly) from program variables. We write such *specifications* as $[pre, post]$. In the style of [17] a program P *satisfies* such a specification iff

$$pre \;\Rightarrow\; wp(P, post) \qquad (2)$$

Paraphrasing [17, p. 16], we say that

> *pre* characterises a set of initial states such that activation of the mechanism P in any one of them will certainly result in a properly terminating happening leaving the system in a final state satisfying *post*.

But we adopt a different style [44] (similarly [5], [50]), writing more directly but equivalently

$$[pre, post] \sqsubseteq P \qquad (3)$$

This we read "the specification $[pre, post]$ is satisfied by P". And we make specifications "first-class citizens", giving their semantics in the same way as all other programming constructs are defined in [17].

Definition 1 *Let* pre, post, *and* R *be predicates over the program variables* v. *We define the weakest pre-condition of the specification* $[pre, post]$ *with respect to the post-condition R as follows:*

$$wp([pre, post], R) \ = \ pre \land (\forall v.post \Rightarrow R)$$

♡

In that definition and below, single letters v refer to a vector of variables (possibly singleton). Definition 1 is discussed in detail in [50] and [44]; the latter allows a more general form in which *post* can refer to the initial state as well.

For the present, we give an informal justification of definition 1: we regard $[pre, post]$ as a statement, and its first component *pre* describes the initial states in which its termination is guaranteed; this is the first conjunct. Its second component *post* describes the final states in which that termination occurs, and so we require also that in all states described by *post* the desired R holds as well; this is the second conjunct.

We now define the relation "is satisfied by" – that is, $\sqsubseteq$ — as in [5], [50], [44], [27]:

Definition 2 *For programs or specifications* P1 *and* P2, *we say that* P1 *is satisfied by* P2, *or equivalently that* P2 *refines* P1, *iff for all post-conditions* R *we have*

$$wp(P1, R) \ \Rightarrow \ wp(P2, R)$$

We write this $P1 \sqsubseteq P2$.

♡

With definitions 1 and 2 we can prove that (2) and (3) are equivalent (see [44]). That equivalence allows us to take $[pre, post]$ as the trivial and most general solution for P in (2). Further, definition 1 agrees with the *Rule of*

We assume below that a and b are fixed.

$$[b^2 - 4ac \geq 0, ax^2 + bx + c = 0]$$

$\sqsubseteq \quad \left[b^2 - 4ac \geq 0,\ x = \frac{-b \pm \sqrt{b^2-4ac}}{2a}\right]$ (standard mathematics)

$\sqsubseteq \quad [b^2 - 4ac \geq 0,\ x^2 = b^2 - 4ac];$ (sequential composition)
$\quad\quad x := (x - b)/2a$

$\sqsubseteq \quad$ **procedure** $Sqrt \;\hat{=}\; [b^2 - 4ac \geq 0,\ x^2 = b^2 - 4ac];$ (copy rule)
$\quad\quad Sqrt;$
$\quad\quad x := (x - b)/2a$

Figure 1: Development of quadratic-solver

adaptation [29] and with the procedure call rule [21, 12.2.1] in the special case where the abstraction is in fact a procedure.

But we are not necessarily linking procedure call and procedural abstraction: procedure call is useful even when the procedure body is executable code; and procedural abstraction is useful even if the implementation ultimately is "inline". Consider the example of figure 1, in which we introduce a parameterless procedure *Sqrt*. There we use specifications [*pre* , *post*] as fully-fledged program constructs, as indeed definition 1 allows us to do.

The conclusion of this exercise would be to refine the remaining specification, but the fact that it is the body of a procedure is now irrelevant:

$$[b^2 - 4ac \geq 0, x^2 = b^2 - 4ac]$$

$\sqsubseteq \quad x := \sqrt{b^2 - 4ac}$

Thus we see that by allowing procedural abstractions — specifications — to mingle with ordinary program constructs, we can with the copy rule accommodate calls to procedures for which we do not yet have the executable code. The specification itself is the text we copy, and definition 1 gives meaning to the result.

4 Parameters

Parameters are used to adapt a general program fragment to a particular purpose — whether or not that fragment is a procedure. Historically, procedures and parametrization are closely linked, and parameter *passing* means "parametrizing a procedure call".

Apparently the simplest example of parametrization is ordinary textual substitution. When substituting into *programs*, much the same rules apply as for substitution into formulae: only *global* (compare *free*) occurrences of x are affected; and capture of l must be avoided by systematic renaming of *local* (compare *bound*) variables. And if we are replacing a variable by a *term*, then that variable cannot appear on the left of :=.

In example (1), we could use substitution to write instead

$\quad\quad y := 0;$
$\quad\quad y := y + 1;$
$\quad\quad \textbf{write } y$

$=\quad y := 0;\quad\quad\quad\quad\quad\quad$ (parametrization)
$\quad\quad (x := x + 1)[x \backslash y];$
$\quad\quad \textbf{write } y$

$=\quad \textbf{procedure } Inc \;\widehat{=}\; x := x + 1;\quad$ (copy rule)
$\quad\quad y := 0;$
$\quad\quad Inc[x \backslash y];$
$\quad\quad \textbf{write } y$

In the final step above, the substitution suggests — intentionally — supplying an actual parameter y for a formal parameter x in the call of procedure *Inc*. But in the previous step, we see $[x \backslash y]$ as a simple substitution.

That style of parametrization, known as *call by name*, is unfortunately not as simple as it appears. Not only is it difficult to implement (requiring "thunks"), but it can be difficult to reason about, as well. If the actual parameters passed lead to distinct names within the procedure for the *same* variable, then the parametrization may lose the crucial property of *monotonicity:* we won't have that $P1 \sqsubseteq P2$ implies $P1[x \backslash T] \sqsubseteq P2[x \backslash T]$.

That phenomenon is known as *aliasing*, and is traditionally associated with procedure call; writers on program development advise us to avoid it. Because of aliasing, call by name (and similarly call by reference: Pascal's **var**) must be used with care. But, in fact, aliasing loses monotonicity — and *that* is why we should avoid it. We can separate the problem from procedure call.

Below we show by example that aliasing loses even equality (trivially, monotonicity also): we have

$\quad\quad (x := 0;\; y := 1) \quad = \quad (y := 1;\; x := 0)$

but

$$\begin{aligned}
&(x := 0;\ y := 1)[x\backslash y] \\
=\ &y := 0;\ y := 1 \\
=\ &y := 1 \\
\neq\ &y := 0 \\
=\ &y := 1;\ y := 0 \\
=\ &(y := 1;\ x := 0)[x\backslash y]
\end{aligned}$$

In the following sections, we define "substitution by value", "by result", and "by value/result"; and we prove that, unlike simple substitution, they are monotonic.

4.1 Substitution by value

For any program P, we write the substitution *by value* in P of term T for variable x as follows:

$P[\textbf{value}\ x\backslash T]$

For simplicity in the following sections, we use the notation $P < R >$ for $wp(P, R)$ (following [27]).

Definition 3 *Substitution by value: if* x *does not occur free in* R, *then*

$P[\textbf{value}\ x\backslash T] < R >\ \ \hat{=}\ \ P < R > [x\backslash T]$

♡

Note that the substitution on the right above is ordinary substitution into the predicate $P < R >$: the weakest precondition is calculated first, then the substitution is made. That convention applies everywhere below.

Substitution by value can be implemented with the well-known *call* by value technique of assignment to an anonymous local variable. It is easily shown that for any program P, variable x, term T, and fresh local variable l, we have

$P[\textbf{value}\ x\backslash T]$

$\begin{aligned}=\ &\textbf{begin var}\ l; \\ &\quad l := T; \\ &\quad P[x\backslash l] \\ &\textbf{end}\end{aligned}$

That implementation, by using ordinary substitution only in a restricted way, avoids the problems we encountered above. First, since the variables l are fresh and distinct, there is no aliasing; second, since the replacing expressions are variables rather than general terms, there is no difficulty when the replaced variables occur on the left of :=.

But our main interest is in monotonicity:

Theorem 1 *Substitution by value is monotonic: if $P \sqsubseteq Q$ then*

$$P[\text{value } x \backslash T] \sqsubseteq Q[\text{value } x \backslash T]$$

Proof: Immediate from definitions 3, 1 and the monotonicity (over $\Rightarrow$) of substitution into predicates: if for predicates X and Y we have $X \Rightarrow Y$, then for any variable v and term T we have also $X[v \backslash T] \Rightarrow Y[v \backslash T]$.
♡

4.2 Substitution by result

For any program P, we write substitution *by result* in P of variable y for variable x as follows:

$$P[\text{result } x \backslash y]$$

This is a more restricted form of substitution than substitution by value, because we substitute a variable y rather than a term T. It is defined as follows:

Definition 4 *Substitution by result: if x does not occur free in R, then*

$$P[\text{result } x \backslash y] < R > \; \widehat{=} \; (\forall x \bullet P < R[y \backslash x] >)$$

♡

Substitution by result can be implemented by the call by result technique of assignment from an anonymous local variable. It can be shown that for any program P, variable x, term T, and fresh local variable l, we have

$$P[\text{result } x \backslash y]$$
$$= \begin{array}{l} \textbf{begin var } l; \\ \quad P[x \backslash l]; \\ \quad y := l \\ \textbf{end} \end{array}$$

For monotonicity, we have

Theorem 2 *Substitution by result is monotonic: if $P \sqsubseteq Q$ then*

$$P[\text{result } x\backslash y] \quad \sqsubseteq \quad Q[\text{result } x\backslash y]$$

Proof: Immediate from definition 4, as for theorem 1.

♡

4.3 Substitution by value/result

For any program P, we write the substitution *by value/result* in P of term y for variable x as follows:

$$P[\text{value result } x\backslash y]$$

Substitution by value/result is a combination of the two substitutions above, and is well-behaved in the same way. We have

Definition 5 *Substitution by value/result: if x does not occur free in R, then*

$$P[\text{value result } x\backslash y] < R > \quad \widehat{=} \quad P < R[y\backslash x] > [x\backslash y]$$

♡

Theorem 3 *Substitution by value/result is monotonic: if $P \sqsubseteq Q$ then*

$$P[\text{value result } x\backslash y] \quad \sqsubseteq \quad Q[\text{value result } x\backslash y]$$

Proof: Immediate from definition 5, as for theorem 2.

♡

The equivalent program fragment is given by

$$P[\text{value result } x\backslash T]$$
$$= \begin{array}{l} \textbf{begin var } l; \\ \quad l := y; \\ \quad P[x\backslash l]; \\ \quad y := l \\ \textbf{end} \end{array}$$

4.4 Apparent limitations

Each of the definitions 3, 4, 5 contains the limitation "if x does not occur free in R". Thus with them we cannot calculate

$$(y := x)[\text{value } x\backslash 0] < x = 0 > \tag{4}$$

It's clear that the weakest precondition in (4) above should be $x = 0$. But calculation (using definition 3 erroneously) reveals instead

$$\begin{aligned}
&(y := x)[\text{value } x\backslash 0] < x = 0 > \\
= \ &(y := x) < x = 0 > [x\backslash 0] \\
= \ &(x = 0)[x\backslash 0] \\
= \ &(0 = 0) \\
= \ &true
\end{aligned}$$

We avoid such problems by extending definitions 3–5 uniformly.

Definition 6 *If the substitution type* **sub** *is* **value**, **result**, *or* **value result**, *we have*

$$P[\text{sub } x\backslash T] < R > \ \hat{=} \ P[x\backslash l][\text{sub } l\backslash T] < R >$$

where l *is a fresh variable, not appearing in* P, T, x, *or* R.

♡

The monotonicity properties persist, and for (4) we now have

$$\begin{aligned}
&(y := x)[\text{value } x\backslash 0] < x = 0 > \\
= \ &(y := l)[\text{value } l\backslash 0] < x = 0 > \\
= \ &(y := l) < x = 0 > [l\backslash 0] \\
= \ &(x = 0)[l\backslash 0] \\
= \ &(x = 0)
\end{aligned}$$

A second limitation is that we have not treated *multiple* parametrization. For example, we cannot calculate

$$(y := x + 1)[\text{value } x, \text{result } y\backslash z, z] \tag{5}$$

We use the normal notation for multiple substitutions: in the above, z replaces x by value and y by result.

We proceed as for simple (multiple) substitutions: for formula P, distinct variables x, y, and terms T, U we know that

$$P[x, y \backslash T, U] \;=\; P[x\backslash l][y\backslash m][l\backslash T][m\backslash U]$$

← for fresh variables l and m. Our definition is therefore

Definition 7 *For any substitution types* **sub1** *and* **sub2**, *distinct variables* x *and* y, *and terms* T, U *we have*

$$P[\mathbf{sub1}\ x, \mathbf{sub2}\ y\backslash T, U] \;\hat{=}\; P[x\backslash l][y\backslash m][\mathbf{sub1}\ l\backslash T][\mathbf{sub2}\ m\backslash U]$$

← *where* l, m *are fresh variables.*
♡

The definition is easily generalised to more than two simultaneous substitutions. In (5) above, we now proceed

$$\begin{aligned}
& (y := x+1)[\mathbf{value}\ x, \mathbf{result}\ y\backslash z, z] < R > \\
=\ & (m := l+1)[\mathbf{value}\ l\backslash z][\mathbf{result}\ m\backslash z] < R > \\
=\ & (\forall m.\, (m := l+1)[\mathbf{value}\ l\backslash z] < R[z\backslash m] >) \\
=\ & (\forall m.\, (m := l+1) < R[z\backslash m][l\backslash z] >) \\
=\ & (\forall m.\, R[z\backslash m][m\backslash l+1][l\backslash z]) \\
=\ & R[z\backslash z+1] \\
=\ & z := z+1 < R >
\end{aligned}$$

Hence the program fragment increments z, as expected.

4.5 Real limitations

Unfortunately, we cannot treat the general cases of "substitution by name" or even "substitution by var". As we have seen, simple substitution (*i.e.*, by name) does not respect equality of programs modulo *wp* unless severe restrictions are made on its use. Those very restrictions, whatever they might be[1], are necessary to achieve monotonicity and can be studied as such. With monotonicity, they can be treated as were the substitutions in section 4 above.

Finally, note that in multiple result parametrization an apparent aliasing can occur if two actual parameters are the same, as in [**result** $x, y\backslash z, z$]. The effect of this must agree with that of multiple assignments $z, z := x, y$ and multiple simple substitutions $[z, z\backslash x, y]$: usually, they are considered syntactically invalid.

---

[1] They vary from writer to writer.

5 Conclusion

Rules for parametrized procedural abstraction are complex. We have argued that they are simplified by considering parametrization, procedure call, and specification separately. The result is a more uniform and orthogonal treatment, in which difficulties are properly located: aliasing for example shown to be a non-monotonic construction.

Combined rules, such as those of [21] and [29], can be derived from ours. It is the program developer's choice whether to use them, or the more basic rules here, or perhaps some other combination especially relevant to his problem.

The separation we have achieved relies essentially on the embedding of specifications within programs: only this allows ALGOL's copy rule to give the meaning of procedure calls independently of the level of abstraction in the procedure body.

We have not treated the call-by-name and call-by-reference parameter passing techniques because they do not fit easily into the standard axiomatic framework of [28] and [17]. In [57, pp. 160-161], for example, call-by-name is treated in a slightly augmented logic in which one can state as a precondition that aliasing is not to occur. That shortcoming of the standard approach, however, we separate from procedures; as we have shown, the real problem is that in general

$$P =_{wp} Q \quad \not\Rightarrow \quad P[x \backslash T] =_{wp} Q[x \backslash T].$$

That is, equality as predicate transformers "$=_{wp}$" is too coarse for these substitutions.

6 Acknowledgements

The work here depends on the original ideas of Hoare [29] and Gries and Levin [22] for the axiomatic treatment of procedure parameters. I believe Back [5] to have been the first to generalise wp in a way similar to ours [44]. He uses single predicates, however, rather than pairs as we do, thus foregoing the advantage of miracles [48].

I am grateful for the very thorough comments of the referees.

Data Refinement by Miracles

Carroll Morgan

Abstract

Data refinement is the transformation in a computer program of one data type to another. Usually, we call the original data type "abstract", and the final data type "concrete". The concrete data type is said to *represent* the abstract.

In spite of recent advances, there remain obvious data refinements that are difficult to prove. We give such a refinement; and we present a new technique that avoids the difficulty.

Our innovation is the use of program fragments that do not satisfy Dijkstra's *Law of the excluded miracle*. These of course can never be implemented, and so they must be eliminated before the final program is reached. But in the intermediate stages of development, they simplify the calculations.

1 Introduction

Data refinement is increasingly becoming a central feature of the modern programming method. Although it is a long-established technique, well-explained for example in [33], it is still developing. Recently it has been extended ([56], [26], [23] and elsewhere) to allow a larger class of refinements than had before been thought desirable. In this paper, we extend it slightly further.

Given two program fragments A and C, we say that C *refines* A iff: C terminates whenever A does; and every result of C is also a possible result of A. In many cases the abstract fragment is a block (or module) using some local (or hidden) variable a, say, and the concrete fragment is to use c instead. The technique of *data* refinement allows the algorithmic structure of the abstract fragment to be carried over into the concrete fragment: that is, if we apply data refinement to the *components* of the abstract fragment, replacing them one-by-one with concrete components, then the concrete fragment refines the abstract fragment *overall*.

Appeared in *Inf. Proc. Lett. 26(5)* (Jan. 1988).
© Copyright 1988, Elsevier Science Publishers B.V. (North-Holland), reprinted with permission.

```
var   bag : bag of Integer;
      sum : Integer;
      summed : bag of Integer;

sum, summed := 0, ≺≻ ;
do summed ≠ bag →
   |[ var x: Integer;
        x: ∈ (bag − summed);
        sum := sum + x;
        summed := summed+ ≺ x ≻
   ]|
od
```

Figure 1: Summing a bag of integers

We exhibit an "obvious" refinement, in which the abstract algorithmic structure is reproduced in the concrete program, but whose corresponding components cannot be data-refined using existing techniques. The inadequacy is due to the required data refinement's being valid only in certain conditions rather than universally. Furthermore, these conditions cannot be expressed in terms of the abstract variables.

Dijkstra's law of the excluded miracle [17, p. 18] states

For all programs P, $wp(P, false) = false$.

Recent work has suggested that derivation of programs is simplified if *miracles* — statements not satisfying the above — are allowed in the intermediate development steps ([50], [2], [44], [54]). We demonstrate a specific application of miracles by showing that they allow conditional data refinement to proceed even when the condition involves concrete variables. The price paid is that some reasoning is then necessary at the concrete level so that the miracles — which can never be executed — are eliminated.

We use the weakest precondition calculus of Dijkstra ([17], [21]) and its associated programming language.

2   An abstract program

Figure 1 contains a program for summing a bag of integers. Bag comprehensions are indicated by "≺ ··· ≻", and "+" is bag (as well as integer) addition. We assume also that variable bag has size N. The statement $x: \in b$ stores in the variable x an arbitrary element of bag b; it is defined

$$wp(x: \in b, R) \;\;\widehat{=}\;\; (b \neq \prec\succ) \wedge (\forall x : x \in b : R)$$

3 A difficult data refinement

We now transform the abstract program of figure 1 into a concrete program, replacing the bags *bag* and *summed* by an array a and an integer n. An *abstraction invariant* will provide the link between the two; it is

$$
\begin{aligned}
& 0 \leq n \leq N \\
\wedge\ & bag = \prec i : i \in 0..N-1 : a[i] \succ \\
\wedge\ & summed = \prec i : i \in 0..n-1 : a[i] \succ
\end{aligned}
\qquad (I)
$$

We now use the formulation from [23] for proving a data refinement correct, paraphrased below:

> An abstract fragment A is data-refined by a concrete fragment $\vec{c}$ under abstraction invariant I iff the following holds:
>
> $$ I \wedge wp(A,\ true) \;\Rightarrow\; wp(\vec{c},\ \neg wp(A,\ \neg I)) \qquad (1) $$

With (1) and our chosen abstraction invariant I, we can show the following data refinements:

Abstract	*Concrete*
$summed := \prec \succ$	$n := 0$
$summed \neq bag$	$n \neq N$
$x :\in (bag - summed)$	$x := a[n]$

But we cannot data-refine $summed := summed + \prec x \succ$, for to do so we would need C satisfying

$$
\begin{aligned}
& I \wedge wp(summed := summed + \prec x \succ,\ true) \\
\Rightarrow\ & wp(C,\ \neg wp(summed := summed + \prec x \succ,\ \neg I))
\end{aligned}
\qquad (2)
$$

It can be shown that *no* assignment to n satisfies (2); in particular, $C =$ "$n := n + 1$" does not.

4 Miraculous programs

We introduce the *guarded command* as follows, for condition G and statement S:

$$ wp(G \rightarrow S,\ R) \;\widehat{=}\; G \Rightarrow wp(S,\ R) $$

Guarded commands can never[1] be implemented by real programs, because they violate Dijkstra's law. Like complex numbers, they can appear during calculation, but must be eliminated if a real (compare implemented) solution is to be reached. The worst offender is $false \rightarrow \mathbf{skip}$, because we have $wp(false \rightarrow \mathbf{skip}, R) = true$, for any R.

Nevertheless, the following statement does satisfy requirement (2):

$$(n \neq N \wedge a[n] = x) \longrightarrow n := n + 1 \qquad (3)$$

5 Eliminating miracles

Although guarded commands violate the law of the excluded miracle, they do obey other laws (distributivity of conjunction, for example). In particular, we have the following:

Law 1 Assignment can be distributed through guarding:

$$\begin{aligned} & v := exp; \ B \rightarrow S \\ = \ & \neg wp(v := exp, \neg B) \longrightarrow v := exp; S \end{aligned}$$

Proof: For all postconditions R, we have

$$\begin{aligned} & wp(\text{``} v := exp; B \rightarrow S\text{''}, R) \\ = \ & def(exp) \wedge (B \Rightarrow wp(S, R))[v\backslash exp] \\ = \ & def(exp) \wedge (B[v\backslash exp] \Rightarrow wp(S, R)[v\backslash exp]) \\ = \ & (def(exp) \Rightarrow B[v\backslash exp]) \Rightarrow (def(exp) \wedge wp(S, R)[v\backslash exp]) \\ = \ & wp(\neg wp(v := exp, \neg B) \rightarrow v := exp; S, R) \end{aligned}$$

(end of law)

With law 1, we can eliminate the miracle; we have in the concrete loop body:

$$\begin{aligned} & x := a[n]; \\ & sum := sum + x; \\ & (n \neq N \wedge a[n] = x) \longrightarrow n := n + 1 \end{aligned}$$

But this equals

$$\begin{aligned} & x := a[n]; \\ & (n \neq N \wedge a[n] = x) \longrightarrow \ & sum := sum + x; \\ & & n := n + 1 \end{aligned}$$

---

[1] ...well, hardly ever: only when the guard is *true*.

```
            ┌─────────────────────────────────────────┐
            │   var  a : array [0..N −1] of Integer;  │
            │        sum : Integer;                   │
            │        n : [0..N];                      │
            │                                         │
            │   sum, n := 0, 0;                       │
            │   do n ≠ N →                            │
            │       |[ var x: Integer;                │
            │          x := a[n];                     │
            │          sum := sum + x;                │
            │          n := n + 1                     │
            │       ]|                                │
            │   od                                    │
            └─────────────────────────────────────────┘
```

Figure 2: Summing an array of integers

For our final step, we note that

$$\begin{aligned}
&\neg wp(x := a[n], \neg(n \neq N \wedge a[n] = x)) \\
=\ &\neg(0 \leq n < N \wedge \neg(n \neq N \wedge a[n] = a[n])) \\
=\ &true
\end{aligned}$$

With this, and law 1 again, we reach the promising

$$\begin{aligned}
true \rightarrow\quad &x := a[n]; \\
&sum := sum + x; \\
&n := n + 1
\end{aligned}$$

But $(true \rightarrow S) = S$ for all S (another law), and so the guard can be eliminated altogether. We are left with the concrete program of figure 2.

6 Conclusion

In our example, a proof of correctness of the concrete operation $n := n + 1$ requires the precondition $a[n] = x$, which cannot be expressed solely in terms of the abstract variables. Hence the proof method of [23] cannot be used. Allowing concrete variables in the precondition is not the solution, for that would destroy our ability to reason at the abstract level independently of possible representations.

It is possible to rearrange our example program, and then to data-refine as a whole the compound statement

$$\begin{aligned}
&x :\in (bag - summed); \\
&summed := summed + \prec x \succ
\end{aligned}$$

In this case concrete variables in preconditions are no longer necessary. But then those two statements must *always* appear adjacent: a severe restriction. We would have lost the important technique of distributing data refinement through program structure.

Guarded commands are useful also when stating rules for algorithmic refinement, in many cases making them simpler by widening their applicability. Mistaken refinements — normally prevented by failure of an applicability condition — instead are allowed to proceed, but generate "infeasible" programs from which the guards cannot be eliminated. The disadvantage of this is that such mistakes can go long unnoticed; but the overwhelming advantage is the decreased proof obligation faced by the developer [48].

Guarded commands were first introduced by Dijkstra [17], who used them only within alternation and iteration constructs. As explained in [44], we "discovered" miracles while extending Dijkstra's language to accommodate embedded specifications.

7 Acknowledgements

The connection between infeasible specifications and guarded commands was pointed out by Tony Hoare, who together with He Jifeng and Jeff Sanders demonstrates in [32] that the standard relational model of programs can give a very elegant formulation of data refinement. They show easily that data refinement is a correctness-preserving technique and that it distributes through the ordinary program constructors. Infeasibility occurs naturally within their work as relations whose domains are partial.

The work reported here falls within the larger context of joint research with Jean-Raymond Abrial [2], Paul Gardiner, Mike Spivey, and Trev Vickers; I am grateful to British Petroleum for supporting our collaboration.

I am grateful also to the painstaking and perceptive referees, and to Jean-Raymond Abrial, Paul Gardiner, David Gries, and Jeff Sanders. Their suggestions have improved the paper significantly.

Auxiliary Variables in Data Refinement

Carroll Morgan

Abstract

A set of local variables in a program is auxiliary if its members occur only in assignments to members of the same set. Data refinement transforms a program, replacing one set of local variables by another set, in order to move towards a more efficient representation of data.

Most techniques of data refinement give a direct transformation. But there is an indirect technique, using auxiliary variables, that proceeds in several stages. Usually, the two techniques are considered separately.

It is shown that the several stages of the indirect technique are themselves special cases of the direct, thus unifying the separate approaches. Removal of auxiliary variables is formalised incidentally.

1 Introduction

Data refinement transforms a program so that certain local variables — called *abstract* — are replaced by other local variables — called *concrete*. Usually the abstract variables range over mathematically abstract values, such as sets, functions *etc.* The concrete variables take values more efficiently represented in a computer, such as arrays.

There are many formalisations of data refinement, all more or less equally powerful. In each a rule is given for producing the concrete statements that correspond to given abstract ones. We call such methods *direct*.

An indirect but equally effective approach uses auxiliary variables. First, the concrete variables are introduced *in parallel* with the abstract variables they ultimately replace. The program is then "massaged" (*i.e., algorithmically* refined) to make those abstract variables auxiliary. Then they are removed.

Our contribution is to show that the auxiliary variable technique is a special case of the direct technique: in fact, it is a composition of direct data refinements. That brings the two styles together, and a more uniform view is gained.

Appeared in *Information Processing Letters 29(6)* (1988).
© Copyright Elsevier Science Publishers B.V. (North Holland), reprinted with permission.

2 The direct technique

Data refinement is described in [30, 33, 23]. An *invariant* is chosen that relates the abstract variables to the concrete variables, and it applies to the whole transformation. Using the invariant, each abstract statement is replaced directly by a concrete statement. We use a recent formulation, taken from [51, 20].

Definition 1 *Direct data refinement:* Let A (C) be the abstract (concrete) statement, a (c) the abstract (concrete) variables, and I the invariant. Then we say that

A is data-refined to C by (I, a, c)

if for all predicates ϕ not containing concrete variables c

$$(\exists a :: I \wedge wp(A, \phi)) \Rightarrow wp(C, (\exists a :: I \wedge \phi)).$$

♡

We assume that the concrete variables c do not appear in the abstract program A.

3 The auxiliary variable technique

This use of auxiliary variables is described in [37], [57, Ch.5], and [17, pp.64–65]. An invariant is chosen, as above. Concrete variables are *added* to the program: their declarations are made in parallel with the existing abstract declarations; and abstract statements are extended so that they maintain the invariant. For example, an abstract assignment $a := AE$, where the expression AE involves abstract variables, is extended to $a, c := AE, E$ by an assignment to the concrete variables. The new expression E may contain variables of either kind, as long as the new statement preserves the invariant I:

$$I \Rightarrow wp(`a, c := AE, E\text{'}, I). \tag{1}$$

In the next step, the program is algorithmically refined to make the abstract variables auxiliary in this sense:

Definition 2 *Auxiliary variables*: A set of local variables is auxiliary if the only executable statements in which its members appear are assignments to members of the same set.
♡

Thus abstract variables must be eliminated from expressions E above and from the guards of alternations and iterations. Finally, the abstract variables are removed from the program entirely; what remains is a data refinement of the original.

4 The correspondence

First, we relate the data refinement (I, a, c) to the composition of two other data refinements.

Lemma 1 *Composition of data refinements:* Let ε be the empty list of variables. If A is data-refined to M by (I, ε, c) and M to C by $(true, a, \varepsilon)$, then A is data-refined to C by (I, a, c).

Proof: Note that empty quantifications ($\exists \varepsilon :: \cdots$) are superfluous. From the assumption and Definition 1, we have for all ϕ not containing c

$$I \wedge wp(A, \phi) \Rightarrow wp(M, I \wedge \phi), \qquad (2)$$

and for all ψ (not containing ε)

$$(\exists a :: wp(M, \psi)) \Rightarrow wp(C, (\exists a :: \psi)). \qquad (3)$$

Now we have for all ϕ not containing c

$$\begin{array}{lll} & (\exists a :: I \wedge wp(A, \phi)) & \\ hence & (\exists a :: wp(M, I \wedge \phi)) & \text{from (2)} \\ hence & wp(C, (\exists a :: I \wedge \phi)) & \text{from (3)} \end{array}$$

That establishes data refinement by (I, a, c).
♡

The correspondence is this: the data refinement (I, ε, c) corresponds to the introduction in parallel of the concrete variables, while preserving I; and the data refinement $(true, a, \varepsilon)$ corresponds to the elimination of the auxiliary variables a. We support that view with the following two lemmas.

Lemma 2 *Introducing concrete variables:* A is data-refined to M by (I, ε, c) if

1. for all ϕ not containing c, $wp(A, \phi) \Rightarrow wp(M, \phi)$; and

2. $I \wedge wp(A, true) \Rightarrow wp(M, I)$.

Proof: For all ϕ not containing c,

	$I \wedge wp(A, \phi)$	
hence	$I \wedge wp(A, \phi) \wedge wp(A, true)$	wp calculus
hence	$wp(M, \phi) \wedge wp(M, I)$	assumptions
hence	$wp(M, I \wedge \phi)$	wp calculus

♡

Assumption 1 of Lemma 2 states that M, over *abstract* variables, is an algorithmic refinement of A. Assumption 2 states that the invariant, linking a and c, is maintained provided A terminates. The example (1) in section 3 is an instance of this.

Lemma 3 *Eliminating auxiliary variables:*
M is data-refined to C by $(true, a, \varepsilon)$ if for all ϕ not containing a,

1. $wp(M, \phi) \Rightarrow wp(C, \phi)$; and

2. $wp(M, \phi)$ contains no a.

Proof: For all ψ

	$(\exists a :: wp(M, \psi))$	
hence	$(\exists a :: wp(M, (\exists a :: \psi)))$	wp calculus
hence	$wp(M, (\exists a :: \psi))$	assumption 2
hence	$wp(C, (\exists a :: \psi))$	assumption 1

♡

Assumption 1 of Lemma 3 states that C, over *concrete* variables, is an algorithmic refinement of M. Assumption 2 states that in M final values of c do not depend on initial values of a — that is, a is auxiliary.

As a final illustration, we apply (3) when a is *not* auxiliary. Taking $c := a$ for M, we must find C such that for all ψ

$$(\exists a :: \psi_{a \Rightarrow c}) \Rightarrow wp(C, (\exists a :: \psi)). \tag{4}$$

But there can be no such C, since

	false	
iff	$wp(C, c = 0 \land c \neq 0)$	excluded miracle
iff	$wp(C, (\exists a :: c = 0)) \land wp(C, (\exists a :: c \neq 0))$	*wp* calculus
iff	$(\exists a :: a = 0) \land (\exists a :: a \neq 0)$	assumption
iff	*true*	

Therefore that data refinement cannot succeed; such failures underlie the soundness of the method.

Note that a variable that *appears* auxiliary in one place might not be auxiliary in another. For example, in $a := a + 1; \cdots; c := a$ we can transform the first statement to **skip** (indeed, that transformation is always possible). But (3) cannot succeed on the second statement: overall, the transformation still fails.

5 Conclusion

We have shown that the two stages of the auxiliary variable technique are data refinements themselves (Lemmas 2, 3), and we have confirmed that the overall result is a data refinement also (Lemma 1). In passing, we have formalised the removal of auxiliary variables.

Data refinement increasingly seems more than a technique for refining data. The transformation $(true, a, \varepsilon)$ — removing auxiliary variables a — has long been used in programming generally. And the transformation (I, ε, c) introduces new variables c which might remain in the program, affording an alternate representation of the structure a.

The transformation $(I, \varepsilon, \varepsilon)$, applied to the exported operations of a module, allows their preconditions to be strengthened (by assuming I); it is successful (compare the failure of (4)) only if each operation establishes I finally. Thus data refinement can also formalise the strengthening of the invariant within a module, though no variables are added or removed. Finally, algorithmic refinement ("massaging") is a special case of that: $(true, \varepsilon, \varepsilon)$.

Definition 1 is slightly more general than [51]: without restriction, we allow free variables in I that do not necessarily appear in a or c. That allows us our empty lists of variables: $(I, \varepsilon, \varepsilon)$ is an extreme example. It also allows invariants that refer to global variables unaffected by the transformation.

We have not discussed the effect of our transformations on guards nor on initialisations. Details are in [51], in [20] where a more theoretical and general approach is taken, and in [47] where it is shown how the transformations allow data refinements to be calculated in practice.

Acknowledgements

I thank Cliff Jones and Edsger W. Dijkstra for the auxiliary variable technique, the former also for reference [37], and Richard Bird, members of IFIP WG 2.3, and the referee for constructive criticism.

Data Refinement of Predicate Transformers

P.H.B. Gardiner and C.C. Morgan

Abstract

Data refinement is the systematic substitution of one data type for another in a program. Usually, the new data type is more efficient than the old, but also more complex; the purpose of the data refinement in that case is to make progress in a program design from more abstract to more concrete formulations.

A particularly simple definition of data refinement is possible when programs are taken to be predicate transformers in the sense of Dijkstra. Central to the definition is a function taking abstract predicates to concrete ones, and this function — a generalisation of the abstraction function — therefore is a predicate transformer as well.

Advantages of the approach are: proofs about data refinement are simplified; more general techniques of data refinement are suggested; and a style of program development is encouraged in which data refinements are calculated directly without proof obligation.

1 Introduction

In many situations, it is more simple to describe the desired result of a task than to describe how a task should be performed. This is particularly true in computer science. Computer programs are very complex, both in their operation and in their representation of information. Yet the task a program performs is often simple to describe.

More confidence in a program's correctness can be gained by describing its intended task in a formal notation. Such *specifications* can then be used as a basis for a provably correct development of the program. The development can be conducted in small steps, thus allowing the unavoidable complexity of the final program to be introduced in manageable pieces.

The process, called *refinement*, by which specifications are transformed into programs has received much study in the past. In particular [30][17][33] have laid down much of the theory and have recognised two forms of refinement.

Firstly, algorithmic refinement: where one makes more explicit the way in which a program operates, usually introducing an algorithm where before there was just a statement of the desired result. And secondly, data refinement: where one changes the structures for storing information, usually replacing some abstract structure that is easily understood, by some more concrete structure that is more efficient.

More recently the emphasis has turned towards providing a uniform theory of program development, in which specifications and programs have equal status. Such a theory is needed to provide the proper setting both for further theoretical work on refinement and for conducting refinement in practice. This goal has been achieved in [5, 50, 44, 48] by extending Dijkstra's language of guarded commands with a *specification statement*. The extended language, by encompassing both programs and specifications, reduces in theory the process of modular program development to program transformation. [50, 44, 48] cover only algorithmic refinement. In this paper we carry on in the same style to include data refinement, and thus give a more complete framework for software development. [51] has made a similar extension.

An important part of our approach is the use of predicate transformers, as in [17], which seem to have several advantages over the relations used in [31]. One is that predicate transformers can represent a form of program conjunction not representable in the relational model. This form of conjunction behaves well under data refinement and can be used to simplify the application of data refinement to specifications. Also, since recursion can be re-expressed in terms of conjunction, this good behaviour allows reasoning about recursion without assuming bounded non-determinism — an unwanted assumption in a theory of programs which includes specifications. But probably the greatest advantage of using predicate transformers is that the theoretical results are so easily applied in practice. In particular, we use a predicate transformer to represent the relationship between abstract and concrete states of a data refinement, and this predicate transformer can be used to calculate directly the concrete program from the abstract program. The calculation maintains the algorithmic structure of the program and adds very little extra complication. Moreover, these calculations do not have any "applicability conditions". No extra proof of correctness is necessary.

2 Predicate transformers

Following [17], we model programs as functions taking predicates to predicates. We blur intentionally the distinction between predicates and the sets of states satisfying them, and therefore we think also of programs as taking sets of (final) states to sets of (initial) states. In any case, for program P and predicate ψ, called the *postcondition*, the application of P to ψ is written $P\,\psi$ and yields a predicate ϕ, called the *weakest precondition* of ψ with respect to P. We say that P *transforms* ψ into ϕ. This predicate ϕ is the weakest one whose truth *initially* guarantees proper termination of P in a state satisfying ψ (finally). The expression $P\psi$ can also be read simply as "P establishes ψ".

The purpose of predicates in the model is to specify sets of states. For this reason, when giving the meaning of a program as a predicate transformer, we will consider only predicates whose free variables are drawn from the program's set of state variables. We will call this set of state variables the program's alphabet (written αP), and call the predicates whose free variables are drawn from a given set of variables x "the predicates on x". Thus, a predicate transformer P can be defined by giving the value of $P\psi$ for all predicates ψ on αP. Of course, we will have to take care not to apply predicate transformers outside their domains.

For clarity, we will sometimes distinguish between program texts and their corresponding predicate transformers, writing $[\![T]\!]$ for the predicate transformer denoted by the program text T.

We define an order $\leq$ on predicates as follows

$$\phi \leq \psi \quad \text{iff} \quad \models \phi \Rightarrow \psi$$

The order $\leq$ permits least upper and greater lower bounds of collections of predicates ϕ_i for which we write respectively

$$\bigvee_i \phi_i \quad \text{and} \quad \bigwedge_i \phi_i.$$

Also the order has a top and bottom

$$\top \text{ and } \bot$$

which correspond to *true* and *false*.

The order on predicates is promoted to predicate transformers in the usual way; for predicate transformers P and Q such that $\alpha P = \alpha Q$:

$$P \sqsubseteq Q \quad \text{iff} \quad \text{for all predicates } \phi \text{ on } \alpha P, \quad P\phi \leq Q\phi$$

This promoted order has least upper and greatest lower bounds as well as a top and bottom element, and they satisfy the following equations:

$$(\bigsqcup_i P_i)\phi = \bigvee_i (P_i\,\phi) \quad (\bigsqcap_i P_i)\phi = \bigwedge_i (P_i\,\phi) \quad \bot \phi = \bot \quad \top \phi = \top$$

All the predicate transformers P we will consider are *monotonic*: for any predicates ϕ and ψ, $\phi \leq \psi$ will imply $P\,\phi \leq P\,\psi$.

3 Algorithmic refinement of predicate transformers

In general, one mechanism is refined by another exactly when every specification satisfied by the first is satisfied also by the second. For predicate transformers we take specifications and satisfaction as follows: a *specification* is a predicate pair $[pre, post]$ comprising the initial assumptions pre and the final requirement $post$; and a program P *satisfies* $[pre, post]$ exactly when

$$pre \Rightarrow P\ post$$

It is now easy to show that P is refined by Q exactly when $P \sqsubseteq Q$.

4 Data refinement of predicate transformers

During algorithmic refinement, *local* variables are usually introduced. And when considering the external behaviour of a program, we ignore the effect it has on its local variables. This gives us a new degree of freedom in refining such programs: we can replace local variables by new ones, so long as the overall effect on the global variables is preserved. This is called *data refinement*.

The following syntax is used to hide (make local) a list of variables x:

$|[\,\mathbf{var}\ x \mid I \bullet P\,]|$

The predicate I states the initialisation of x, and P is the program within which the variables x may be used. This construct is used only if the alphabet of P contains x. The alphabet of the result is that of P with x removed. The meaning of the construct is as follows: for any predicate ψ on $(\alpha P - x)$

$[\![\,|[\,\mathbf{var}\ x \mid I \bullet P\,]|\,]\!]\psi \ \widehat{=}\ (\forall x \bullet I \Rightarrow [\![P]\!]\psi)$

We now define data refinement. Let us suppose we wish to replace the list of variables a (the abstract variables) in

$|[\,\mathbf{var}\ a \mid I \bullet P\,]|$

by some other list of variables c (the concrete variables), and let the variables of αP, other than a, be g (the global variables). We choose any predicate transformer rep that takes predicates on the variables a, g to predicates on the variables c, g. Then for programs P and P', we write $P \preceq P'$ to mean that P is data-refined by P'.

Definition 1 $P \preceq P'$ iff $rep \circ P \sqsubseteq P' \circ rep$
where the operator $\circ$ is functional composition (of predicate transformers).

We will see that for data refinement to be well behaved, we must restrict our choice for rep. We choose rep satisfying the following two properties:

- rep is *monotonic*: $(\forall a \bullet \phi \Rightarrow \psi) \Rightarrow (\forall c \bullet rep\ \phi \Rightarrow rep\ \psi)$;
- rep is $\vee$-*distributive*: $rep\ (\bigvee_i \phi_i) = \bigvee_i (rep\ \phi_i)$

Note that *strictness* is a special case of $\vee$-*distribution* (i.e. $rep\ \bot = \bot$). Note also that this form of monotonicity is stronger than the usual. Further properties that follow from these are proven below. In these proofs and others we will make use of the fact that the two lists of variables a and g are, by definition, disjoint and so the variables a do not occur free in the predicates on g.

Lemma 1 *If ϕ is a predicate on g, then*

$$rep\ \phi \leq \phi$$

proof:

$\neg \phi \Rightarrow (\forall a \bullet \phi \Rightarrow \bot)$	since a is not free in ϕ
$\neg \phi \Rightarrow (rep\ \phi \Rightarrow rep\ \bot)$	monotonicity of rep
$\neg \phi \Rightarrow (rep\ \phi \Rightarrow \bot)$	strictness of rep
$\neg \phi \Rightarrow \neg rep\ \phi$	predicate calculus
$rep\ \phi \Rightarrow \phi$	predicate calculus

♡

Lemma 2 *If ϕ is a predicate on a, g and ψ is a predicate on g, then*

$$(rep\ \phi) \wedge \psi \leq rep(\phi \wedge \psi)$$

proof:

$\psi \Rightarrow (\forall a \bullet \phi \Rightarrow \phi \wedge \psi)$	since a is not free in ψ
$\psi \Rightarrow (rep\ \phi \Rightarrow rep(\phi \wedge \psi))$	monotonicity of rep
$(rep\ \phi) \wedge \psi \Rightarrow rep(\phi \wedge \psi)$	predicate calculus

♡

In subsequent sections we will discuss a particularly convenient choice for rep, and will show how to *calculate* suitable P'. For now, we give the fundamental theorem of data refinement:

Theorem 1 *If for suitable rep (as defined above) we have shown that $P \preceq P'$, then*

$$|[\,\text{var } a \mid I \bullet P\,]| \sqsubseteq |[\,\text{var } c \mid \text{rep } I \bullet P'\,]|$$

proof: Let ψ be any predicate on g, then

$$\begin{aligned}
& [\![\,|[\,\text{var } a \mid I \bullet P\,]|\,]\!]\psi \\
&= (\forall\, a \bullet I \Rightarrow [\![P]\!]\psi) & semantics \\
&\leq (\forall\, c \bullet \text{rep } I \Rightarrow \text{rep }[\![P]\!]\psi) & monotonicity\ of\ rep \\
&\leq (\forall\, c \bullet \text{rep } I \Rightarrow [\![P']\!]\text{rep }\psi) & hypothesis \\
&\leq (\forall\, c \bullet \text{rep } I \Rightarrow [\![P']\!]\psi) & lemma\ 1\ and\ monotonicity\ of\ P' \\
&= [\![\,|[\,\text{var } c \mid \text{rep } I \bullet P'\,]|\,]\!]\psi & semantics
\end{aligned}$$

♡

5 The programming language

The *programming language* is the syntax with which we describe predicate transformers. Here we will use Dijkstra's language [17] with several extensions.

5.1 Extensions

All of the predicate transformers P which can be described by Dijkstra's original language satisfy the following properties:

strictness $P\bot = \bot$;

monotonicity $\phi \leq \psi$ implies $P\,\phi \leq P\,\psi$;

∧-distributivity $P\,(\bigwedge_i \psi_i) = \bigwedge_i (P\,\psi_i)$, for any non-empty family $\{\psi_i\}_i$;

continuity $P\,(\bigvee_i \psi_i) = \bigvee_i (P\,\psi_i)$, for any chain $\{\psi_i\}_i$.

We will see that some of these properties fail in our extended language.

The first extension was given in section 4 above: the introduction of local variables $|[\,\text{var } x \mid I \bullet P\,]|$. It preserves strictness, monotonicity, ∧-distributivity, and continuity.

The second extension is the *specification*. It is written $[pre, post]$ (as in section 2, but here we add it to the *programming* language). The meaning of this construct depends on the alphabet. It is defined for alphabet x as follows:

Definition 2 *For any predicate ψ on x,*

$$[\![\,[pre, post]\,]\!]\psi \;\widehat{=}\; pre \wedge (\forall\, x \bullet post \Rightarrow \psi)$$

Specifications are monotonic and $\wedge$-distributive, but $[\top, \bot]$ is not strict, and $[\top, \top]$ is not continuous (take any chain $\{\psi_i\}_i$ with $\bigvee_i \psi_i = \top$ but $\psi_i \neq \top$ for any i).

The third extension is *conjunction* of programs, written $\ddagger_i P_i$ for any family $\{P_i\}_i$ of programs. Its meaning is given as follows:

Definition 3 $[\![\,\ddagger_i P_i\,]\!] \;\widehat{=}\; \bigsqcup_i [\![P_i]\!]$

Thus the conjunction of a family of programs is the worst program that is better than each member of the family. Conjunction preserves strictness, monotonicity, and continuity, but not $\wedge$-distributivity: consider

$$[\![\,[\top, \psi]\,\ddagger\,[\top, \neg\psi]\,]\!](\psi \wedge \neg\psi)$$

Conjunction is an important counterpart of the specification, since the specification as it stands in definition 2 does not allow the post-condition to refer to the state before execution. Using conjunction and specification together, we can rectify this problem. For example, the assignment statement $n := n + 1$ can be expressed as $\ddagger_i [n = i, n = i + 1]$.

This completes the extension of Dijkstra's language. We can now see that the only property retained from the original language is monotonicity, since each of the other properties is violated by at least one of the program constructors.

5.2 Generalisations

Having accepted the loss of strictness, continuity and $\wedge$-distributivity, we are able to make other generalisations of the language. Choice and guarding need not be restricted to use within the do$\cdots$od and if$\cdots$fi constructs. They can instead be defined as language constructs in their own right.

Definition 4 *Choice: For any family $\{P_i\}$ of programs we define their* choice *as follows:*

$$[\![\,[\!]_i\, s_i\,]\!] \;\widehat{=}\; \bigsqcap_i [\![s_i]\!]$$

Definition 5 *Guarding: For predicate G and program P, we define the guarded command $G \to P$ as follows:*

$$[\![G \to P]\!]\psi \;\;\widehat{=}\;\; G \Rightarrow [\![P]\!]\psi$$

Definition 6 *Recursion:* We must consider program contexts, which denote functions from predicate transformers to predicate transformers. If $\mathcal{C}$ is a program context and P a program, then $\mathcal{C}(P)$ is a program also. The context $\mathcal{C}$ should be thought of as a program structure into which program fragments (for example, P) can be embedded. We have

$$[\![\mu X \bullet \mathcal{C}(X)]\!] \;\;\widehat{=}\;\; \mathbf{fix}[\![\mathcal{C}]\!]$$

where **fix** *takes the least fixed point of a function (from predicate transformers to predicate transformers in this case).*

With these definitions, we can if we wish define the conventional **if** $\cdots$ **fi** and **do** $\cdots$ **od** constructors as appropriate combinations. We have

Definition 7 *Alternation:*

$$\begin{array}{ll} \mathbf{if} & G_1 \to P_1 \\ [\!] & \vdots \\ [\!] & G_n \to P_n \\ \mathbf{fi} & \end{array}$$

is an abbreviation for

$$(\mathop{[\!]}_{i=1}^{n} G_i \to P_i) \,[\!]\, \neg (\bigvee_i G_i) \to \mathbf{abort}$$

Definition 8 *Iteration:*

$$\begin{array}{ll} \mathbf{do} & G_1 \to P_1 \\ [\!] & \vdots \\ [\!] & G_n \to P_n \\ \mathbf{od} & \end{array}$$

is an abbreviation for

$$(\mu X \bullet (\mathop{[\!]}_{i=1}^{n} G_i \to P_i; X) \,[\!]\, \neg (\bigvee_i G_i) \to \mathbf{skip})$$

6 Distribution of data refinement

After theorem 1, the most important property of data refinement is that it distributes through the algorithmic constructors of our programming language. Only then can one carry over the algorithmic structure of the abstract program onto the concrete program. We prove this distribution for each constructor below:

Lemma 3 *Sequential composition: If $P \preceq P'$ and $Q \preceq Q'$ then $P; Q \preceq P'; Q'$*

proof:

$rep \circ [\![P; Q]\!]$
$\quad = rep \circ [\![P]\!] \circ [\![Q]\!]$ \hfill *semantics*
$\quad \sqsubseteq [\![P']\!] \circ rep \circ [\![Q]\!]$ \hfill *hypothesis*
$\quad \sqsubseteq [\![P']\!] \circ [\![Q']\!] \circ rep$ \hfill *hypothesis and monotonicity of P'*
$\quad = [\![P'; Q']\!] \circ rep$ \hfill *semantics*

♡

Lemma 4 *Skip:* skip $\preceq$ skip

proof:

$rep \circ [\![skip]\!]$
$\quad = rep \circ Id$ \hfill *semantics*
$\quad = rep$ \hfill *property of Id*
$\quad = Id \circ rep$ \hfill *property of Id*
$\quad = [\![skip]\!] \circ rep$ \hfill *semantics*

♡

Lemma 5 *Abort:* abort $\preceq$ abort

proof:

$rep \circ [\![abort]\!]$
$\quad = rep \circ \bot$ \hfill *semantics*
$\quad = \bot$ \hfill *strictness of rep*
$\quad = \bot \circ rep$ \hfill *property of $\bot$*
$\quad = [\![abort]\!] \circ rep$ \hfill *semantics*

♡

Lemma 6 *Guarding: To deal with guarded commands we will need another function from abstract predicates to concrete predicates.*

$\overline{rep}\ \psi \mathrel{\widehat{=}} \neg\,(rep\,\neg\,\psi)$

We then have the following result for guarded commands. If $P \preceq P'$ then $(G \to P) \preceq ((\overline{rep}\ G) \to P')$

proof:

$rep\ [\![G \to P]\!]\ \psi$
$= rep\ (G \Rightarrow [\![P]\!]\ \psi)$ *semantics*
$= rep\ (\neg\ G \vee [\![P]\!]\ \psi)$ *predicate calculus*
$= (rep\ \neg\ G) \vee (rep\ [\![P]\!]\ \psi)$ $\vee$*-distributivity of rep*
$= \neg\ (\overline{rep}\ G) \vee (rep\ [\![P]\!]\ \psi)$ *definition of* $\overline{rep}$
$= (\overline{rep}\ G) \Rightarrow (rep\ [\![P]\!]\ \psi)$ *predicate calculus*
$\leq (\overline{rep}\ G) \Rightarrow ([\![P']\!]\ rep\ \psi)$ *hypothesis*
$= [\![(\overline{rep}\ G) \to P']\!]\ rep\ \psi$ *semantics*

♡

Lemma 7 *Choice: If for each i $P_i \preceq P_i'$ then $[\!]_i P_i \preceq [\!]_i P_i'$*

proof:

$rep \circ [\![\ [\!]_i P_i\ ]\!]$
$= rep \circ (\sqcap_i [\![P_i]\!])$ *semantics*
$\sqsubseteq \sqcap_i (rep \circ [\![P_i]\!])$ *monotonicity of rep*
$\sqsubseteq \sqcap_i ([\![P_i']\!] \circ rep)$ *hypothesis*
$= (\sqcap_i [\![P_i']\!]) \circ rep$ *property of* $\sqcap$
$= [\![\ [\!]_i P_i'\ ]\!] \circ rep$ *semantics*

♡

Lemma 8 *Conjunction: If for each i $P_i \preceq P_i'$ then $\ddagger_i P_i \preceq \ddagger_i P_i'$*

proof:

$rep \circ [\![\ddagger_i P_i]\!]$
$= rep \circ (\sqcup_i [\![P_i]\!])$ *semantics*
$= \sqcup_i (rep \circ [\![P_i]\!])$ $\vee$*-distributivity of rep*
$\sqsubseteq \sqcup_i ([\![P_i']\!] \circ rep)$ *hypothesis*
$= (\sqcup_i [\![P_i']\!]) \circ rep$ *property of* $\sqcup$
$= [\![\ddagger_i P_i]\!] \circ rep$ *semantics*

♡

Lemma 9 *Recursion: We first promote data refinement to program contexts: we say that $C \preceq C'$ exactly when for all pairs of programs P and P' such that $P \preceq P'$, we have $C(P) \preceq C'(P')$ as well. The result for recursion is then as follows:*

If $C \preceq C'$, then

$$(\mu X \bullet C(X)) \preceq (\mu X \bullet C'(X))$$

proof: The Knaster-Tarski theorem asserts the existence of an ordinal γ such that **fix** $F = F^\gamma \perp$, where

$$F^0 \ X = X \quad F^{\alpha+1} \ X = F(F^\alpha \ X) \quad F^\gamma \ X = \bigsqcup_{\beta < \gamma} (F^\beta \ X)$$

Hence, it is sufficient to prove $[\![C]\!]^\gamma \perp \preceq [\![C']\!]^\gamma \perp$ for all γ. This can be proven by induction. The base case follows from Lemma 5, the step case follows from $C \preceq C'$ and the limit case follows from Lemma 8. ♡

7 Data refinement of specifications

In the preceding section, we showed that data refinement can be performed piecewise (a term we borrow from [51]), thus maintaining the algorithmic structure of a program. We now consider the pieces lying within the structure.

There are two constructs to consider, the specification and the assignment. In fact, we can ignore the assignment statement since it is readily transformed into a simple specification.

The following theorems provide a method for calculating the data refinement of specifications, and show that this method produces the most general data refinement.

In both theorems we will, again, write a for the list of abstract variables, c for the list of concrete variables, g for the remaining (global) variables and rep for the representation predicate transformer.

Theorem 2 $[pre, post] \preceq [rep \ pre, rep \ post]$
proof: Let ψ be any predicate on g, a, then

$$\begin{aligned}
& rep \ [\![\, [pre, post] \,]\!] \ \psi \\
&= rep \ (pre \wedge (\forall g, a \bullet post \Rightarrow \psi)) & \text{semantics} \\
&\leq (rep \ pre) \wedge rep \ (\forall g, a \bullet post \Rightarrow \psi) & \text{monotonicity of } rep \\
&\leq (rep \ pre) \wedge (\forall g, a \bullet post \Rightarrow \psi) & \text{lemma 1} \\
&\leq (rep \ pre) \wedge (\forall g, c \bullet rep \ post \Rightarrow rep \ \psi) & \text{monotonicity of } rep \\
&= [\![\, [rep \ pre, rep \ post] \,]\!] \ rep \ \psi & \text{semantics}
\end{aligned}$$

♡

Theorem 3 *If* $[pre, post] \preceq P$ *then* $[rep\ pre, rep\ post] \sqsubseteq P$.

proof: Let ψ be any predicate on g, c, then

$$
\begin{aligned}
&\llbracket\ [rep\ pre, rep\ post]\ \rrbracket \psi \\
&= (rep\ pre) \wedge (\forall g, c \bullet rep\ post \Rightarrow \psi) &&\text{semantics}\\
&\leq (rep\ pre) \wedge (\forall g, a \bullet post \Rightarrow \bigvee_{\forall g, c \bullet rep\ x \Rightarrow \psi} x) &&\text{properties of } \vee\\
&\leq rep\ (pre \wedge (\forall g, a \bullet post \Rightarrow \bigvee_{\forall g, c \bullet rep\ x \Rightarrow \psi} x)) &&\text{lemma 2}\\
&= rep\ (\llbracket\ [pre, post]\ \rrbracket \bigvee_{\forall g, c \bullet rep\ x \Rightarrow \psi} x) &&\text{semantics}\\
&\leq P\ (rep\ \bigvee_{\forall g, c \bullet rep\ x \Rightarrow \psi} x) &&\text{hypothesis}\\
&= P\ (\bigvee_{\forall g, c \bullet rep\ x \Rightarrow \psi} rep\ x) &&\vee\text{-distributivity of } rep\\
&\leq P\ \psi &&\text{properties of } \vee \text{ and monotonicity of } P
\end{aligned}
$$

♡

8 Data refinement in practice

So far we have given no indication as to how one chooses a suitable representation transformer *rep*. In fact, there may be many classes of program transformation that can be supported by the theory of the proceeding sections. We can, though, cite one example that proves very useful in practice. This definition of *rep*, which is described below, gives the same form of data refinement as that in [23], [51] and also, under the name of downward simulation, in [32].

When performing data refinement, one always intends that the abstract and concrete states should correspond in some way. This correspondence can be expressed as a predicate over the two sets of state variables. From such a predicate (*I* say) the representation transformer can be defined as follows.

$$rep\ \phi \;\widehat{=}\; (\exists a \bullet I \wedge \phi)$$

It is easy to verify that this choice of *rep* has the required properties (i.e. *monotonicity* and $\vee - distributivity$).

$\overline{rep}$ also has a simple form.

$$\overline{rep}\ \phi = (\forall a \bullet I \Rightarrow \phi)$$

Now that we have a definition for *rep* with such a simple form, we can see how easily data refinements can be calculated. The two simple transformers, *rep* and $\overline{rep}$ can be applied directly to the predicates of any abstract program so as to calculate a concrete refinement. In the abstract program all the pre and post conditions of specifications are replaced by their image under *rep* and the guards by their image under $\overline{rep}$ thus leaving the algorithmic structure unchanged. The result is guaranteed correct by the theorems of the previous sections.

9 Conclusions

We have presented the familiar technique of data refinement in the novel context of predicate transformers. In doing so we have drawn on other recent work in program development: the factoring of Dijkstra's language into smaller pieces (Definitions 7 and 8); the use of recursion in practice rather than iteration as the basis for unbounded computations in Dijkstra's language [15]; and the mixing of specification and program [5, 50, 44, 48]. And in Definition 3 we give a further factorisation: with program conjunction, the technique of "logical" variables is formalised. All of this comes together to promote a style of program design in which steps are made by calculation rather than via proof obligations.

It is clear, though, that proof cannot be avoided altogether! In practice, the necessary truths of predicate calculus are drawn on when strengthening post-conditions and weakening preconditions — and virtually nowhere else. In this respect perhaps the proofs have moved rather than disappeared. Their confinement though makes the *other* rules easier to apply in practice. Certainly they are easier to remember: the formulation of data refinement in theorem 2 is simpler than any other of which we are aware.

Using predicate transformers as a model, rather than the relations of [32], affords several advantages. One is that the dependence on continuity is more easily broken. That allowed us to extend the work of [32] so that it applies to a language that includes constructs for specification as well as programming. Another advantage is the ease with which the theoretical results are applied in practice. Both these advantages are related to our conjunction, which can not be represented in the relational model. Conjunction simplifies the expression of specifications in our language; and this, in turn, permits the very simple method of refinement calculation. In contrast, the method of calculation in [32], although theoretically simple, gives rise to very large and unwieldy expressions in practice.

Our relaxing of Dijkstra's "healthiness" conditions has left us only with monotonicity: continuity, strictness, and $\wedge$-distributivity are gone. That is similar to [50], where continuity and strictness are dropped so that the guard and choice symbols can be given meaning as operators in their own right. We too proposed this in [44], but have taken the process further, dropping also $\wedge$-distributivity, so that we can define the conjunction operator which is the key to simplifying the calculation of data refinements.

Our results are potentially more general than those of [51], since we recognise how the abstraction condition is, itself, applied as a predicate transformer, and base all are proofs on two properties of it. By doing this we make the structure of the proofs more explicit, and also leave open the possibility of finding other predicate transformers with these properties which can, therefore, also be used for data refinement.

We also prove several results that [51] does not. Theorem 1 forms the important link between data refinement of local variables and operational refinement. Theorem 3 shows that our method of calculation yields the weakest program that is a data refinement of the original and thus that no loss of choice is incurred by calculation.

10 Acknowledgements

An early description of data refinement appeared in [30], and it later was made part of the Vienna Development Method [33]. Specifications were embedded within programs by [5], who also treated data refinement. The first connection between data refinement and weakest preconditions was made by [5], though it had been earlier presented by [17] as a technique based on auxiliary variables. ([41] explains the connection between these.) Most recently, [51] has given the same formulation as section 8 above, and his work has improved ours in several ways.

Our own work owes much to collaboration with Jean-Raymond Abrial and Mike Spivey who have been a constant source of new and exciting ideas. Much of our contact with other researchers has been made possible by the generosity of British Petroleum Ltd.

Data Refinement by Calculation

Carroll Morgan and P.H.B. Gardiner

Abstract

Data refinement is the systematic substitution of one data type for another in a program. Usually, the new data type is more efficient than the old, but possibly more complex; the purpose of the data refinement in that case is to make progress in program construction from more abstract to more concrete formulations.

A recent trend in program construction is to *calculate* programs from their specifications; that contrasts with proving that a *given* program satisfies some specification. We investigate to what extent the trend can be applied to data refinement.

1 Introduction

In [4], Back proposed an extension of Dijkstra's calculus [17] where specifications and programs are given equal status during program construction. Later interest in specifications generally has led quite recently to further work on such constructions [50, 44, 48, 7, 43, 6]. The style is now known as the *refinement calculus*.

Characteristic of any calculus is that it is used for *calculation*, not just description. The refinement calculus, therefore, should allow programs to be calculated from their specifications. It does indeed allow presentations in which each intermediate design follows from a previous design according to some *law of refinement*. That contrasts with the more well-known style in which intermediate designs are first proposed and then proved to follow from their antecedents. Our hope is that constructions in the refinement calculus will proceed more smoothly, and that proof obligations will be reduced. That is the point of a calculus, and it can be observed elsewhere: for example, in the differential calculus one uses laws of differentiation, not proofs from first principles. For differentiation, the process is now mechanical. In the integral calculus, we have laws too — but there, as in the refinement calculus, success is not guaranteed.

Appeared in *Acta Informatica* 27 (1991).
© Copyright Springer-Verlag 1991.

Data refinement is a special case of refinement: one replaces an abstract type by a more concrete type in a program while preserving its algorithmic structure. Abstract operations are similarly replaced by corresponding concrete operations. It is a well-established technique, with its own specialised proof rules [30, 33].

Our principal contribution is to draw data refinement into the calculational style: we show how to *calculate* data refinements rather than prove them. Our emphasis here is on practice, in contrast to our earlier [20]: this paper gives applications of that theory, though for convenience we have presented afresh some proofs which are corollaries by specialisation of [20]. Recent work by Morris [51] addresses the same concerns that we do.

In passing we formalise *logical constants*, long used in program derivation, but not until now treated rigorously. Their use in programs loses the property of conjunctivity, another of Dijkstra's healthiness laws [17]. (The law of the excluded miracle, and continuity, have already been abandoned [44, 50, 54, 15].)

This work relies on the ideas of the refinement calculus, reviewed in Sections 2 and 3 below. More detail can be found in [44, 48, 50, 5].

2 Refinement

We consider Dijkstra's programming language [17], whose meaning is given by *predicate transformers*. For any program P, we write $[\![P]\!]$ for its *meaning*; and that meaning is a function from (desired) final assertions to (necessary) initial ones:

> For any formula ψ over state variables, and program P, $[\![P]\!]\psi$ is the *weakest* formula whose truth in an initial state ensures that activation of P will lead to a final state in which ψ is true.

Thus we write $[\![P]\!]\psi$ for Dijkstra's $wp(P, \psi)$.

2.1 Algorithmic refinement

A program P is *algorithmically* refined by another P' whenever every specification satisfied by P is satisfied by P' also. We restrict our specifications, however, to formulae $\phi \Rightarrow [\![P]\!]\psi$ which state "the program P must be such that its activation in a state in which ϕ is true will lead to a state in which ψ is true." We do not, for example, specify time or space constraints.

Definition 1 *Algorithmic refinement:* Program P is algorithmically refined by program P' precisely when, for all formulae ϕ and ψ over the program variables,

$$\phi \Rightarrow \llbracket P \rrbracket \psi \quad \text{implies} \quad \phi \Rightarrow \llbracket P' \rrbracket \psi.$$

We write $P \sqsubseteq P'$ for that relationship.
♡

The following is an easy consequence of Definition 1, and is what we will use in practice:

Lemma 1 *Algorithmic refinement:* For programs P and P', we have $P \sqsubseteq P'$ precisely when

$$\llbracket P \rrbracket \psi \Rightarrow \llbracket P' \rrbracket \psi \quad \text{for all formulae } \psi \text{ over the program variables.}$$

Proof: For *if*, note that $\phi \Rightarrow \llbracket P \rrbracket \psi$ and $\llbracket P \rrbracket \psi \Rightarrow \llbracket P' \rrbracket \psi$ imply $\phi \Rightarrow \llbracket P' \rrbracket \psi$ as required; for *only if*, take ϕ to be $\llbracket P \rrbracket \psi$ itself.
♡

We assume that in Lemma 1 we may limit our choice of formulae ψ to those containing only variables free either in P or P' or both.

2.2 Data refinement

Data refinement arises as a special case of algorithmic refinement. A program P is data-refined to another program P' by a transformation in which some so-called *abstract* data-type in P is replaced by a *concrete* data-type in P'. The overall effect is an algorithmic refinement of the block in which the abstract data type is declared.

For that, we add *local variables* to Dijkstra's language in the following (standard) way:

Definition 2 *Local variables:* For (list of) variables l, formula I (the initialisation), and program P, the construction

$$|[\, \mathbf{var}\ l \mid I \bullet P \,]|$$

is a *local block* in which the local variables l are introduced for the use of program P; they are first assigned initial values such that I holds. We define, for ψ not containing l,

$$\llbracket\, |[\, \mathbf{var}\ l \mid I \bullet P \,]|\, \rrbracket \psi \ \widehat{=}\ (\forall l \bullet I \Rightarrow \llbracket P \rrbracket \psi)$$

♡

Note that the scope of quantifiers is indicated explicitly by parentheses $(\forall \cdots)$; the spot • reads "such that".

Where a postcondition ψ does contain the local variable l, Definition 2 can be applied after systematic change of the local l to some fresh l'. We assume therefore that such clashes do not occur.

Where appropriate, we consider *types* to be simply sets of values, and will write $|[\text{ var } l: T \mid I \bullet P]|$ for $|[\text{ var } l \mid (l \in T \wedge I) \bullet P]|$; thus a variable is initialised to some value in its type. And if I is just *true* we may omit it, writing $|[\text{ var } l \bullet P]|$ or $|[\text{ var } l: T \bullet P]|$ as appropriate.

Now data-refinement transforms an abstract block $|[\text{ var } a \mid I \bullet P]|$ to a concrete block $|[\text{ var } c \mid I' \bullet P']|$. We assume that the concrete variables c do not occur in the abstract program I and P, and *vice versa*. The transformation has these characteristics:

1. The concrete block algorithmically refines the abstract block:

   $|[\text{ var } a \mid I \bullet P]| \sqsubseteq |[\text{ var } c \mid I' \bullet P']|.$

2. The abstract variable declarations **var** a are replaced by concrete variable declarations **var** c.

3. The abstract initialisation I is replaced by a concrete initialisation I'.

4. The abstract program P, referring to variables a but not c, is replaced by a concrete program P' referring to variables c but not a; moreover, the algorithmic structure of P is reproduced in P' (see below).

The four characteristics are realised as follows. An *abstraction invariant* AI is chosen which links the abstract variables a and the concrete variables c. It may be *any* formula, but usually will refer to a and c at least. (See Section 4.1 below for a discussion of the impracticality of choosing *false* as the abstraction invariant.) The concrete initialisation I' must be such that $I' \Rightarrow (\exists a \bullet AI \wedge I)$. For the concrete program we define a relation $\preceq$ of data-refinement:

Definition 3 *Data refinement:* A program P is said to be data-refined by another program P', using abstraction invariant AI, abstract variables a and concrete variables c, whenever for all formulae ψ not containing c free we have

$$(\exists a \bullet AI \wedge [\![P]\!]\psi) \Rightarrow [\![P']\!](\exists a \bullet AI \wedge \psi)$$

We write this relation $P \preceq_{AI,a,c} P'$, and omit the subscript $_{AI,a,c}$ when it is understood from context.
♡

Definition 3 is appropriate for two reasons. The first is that it guarantees characteristic 1, as we now show.

Theorem 1 *Soundness of data-refinement:* If $I' \Rightarrow (\exists a \bullet AI \wedge I)$ and $P \preceq P'$, then

$$\| [\text{ var } a \mid I \bullet P]\| \sqsubseteq \| [\text{ var } c \mid I' \bullet P']\|$$

Proof: Consider any ψ not containing a or c free. We have

$$
\begin{aligned}
& [\![\| [\text{ var } a \mid I \bullet P]\|]\!]\psi \\
=\ & (\forall a \bullet I \Rightarrow [\![P]\!]\psi) & \text{Definition 2} \\
=\ & (\forall c, a \bullet I \Rightarrow [\![P]\!]\psi) & c \text{ not free in above} \\
\Rightarrow\ & (\forall c, a \bullet AI \wedge I \Rightarrow AI \wedge [\![P]\!]\psi) \\
\Rightarrow\ & (\forall c \bullet (\exists a \bullet AI \wedge I) \Rightarrow (\exists a \bullet AI \wedge [\![P]\!]\psi)) \\
\Rightarrow\ & (\forall c \bullet I' \Rightarrow (\exists a \bullet AI \wedge [\![P]\!]\psi)) & \text{assumption} \\
\Rightarrow\ & (\forall c \bullet I' \Rightarrow [\![P']\!](\exists a \bullet AI \wedge \psi)) & \text{assumption; Definition 3} \\
\Rightarrow\ & (\forall c \bullet I' \Rightarrow [\![P']\!]\psi) & \text{monotonicity; } a \text{ not free in } \psi \\
=\ & [\![\| [\text{ var } c \mid I' \bullet P']\|]\!]\psi & \text{Definition 2}
\end{aligned}
$$

♡

The second reason our Definition 3 is appropriate is that it distributes through program composition. This is shown in [51, 20], and we refer the reader there for details. Here, for illustration, we treat sequential composition; alternation and iteration are dealt with in Sections 4 and 6 below.

Lemma 2 *Data-refinement distributes through sequential composition:* If $P \preceq P'$ and $Q \preceq Q'$ then $(P; Q) \preceq (P'; Q')$.

Proof: Let ψ be any formula not containing c. Then

$$
\begin{aligned}
& (\exists a \bullet AI \wedge [\![P; Q]\!]\psi) \\
=\ & (\exists a \bullet AI \wedge [\![P]\!]([\![Q]\!]\psi)) & \text{semantics of ``;''} \\
\Rightarrow\ & [\![P']\!](\exists a \bullet AI \wedge [\![Q]\!]\psi) & P \preceq P' \\
\Rightarrow\ & [\![P']\!]([\![Q']\!](\exists a \bullet AI \wedge \psi)) & Q \preceq Q'; \text{monotonicity} \\
=\ & [\![P'; Q']\!](\exists a \bullet AI \wedge \psi) & \text{semantics of ``;''}
\end{aligned}
$$

♡

It is the distributive property illustrated by Lemma 2 that accounts for characteristic 4 above: if for example P is $P_1; P_2; \cdots P_n$ then we can construct P' with $P \preceq P'$ simply by taking $P' = P'_1; P'_2; \cdots P'_n$ with $P_i \preceq P'_i$ for each i. It is in this sense that the structure of P is preserved in P'. We will see in Section 4 below that this carries through for alternations and iterations also.

3 Language extensions

We extend Dijkstra's language in two ways. With the *specification statement* we allow specifications and executable program fragments to be mixed, thus promoting a more uniform development style. With *program conjunction* we make more rigorous the use of so-called *logical constants*, which appear in specifications but not in executable programs.

3.1 Specification statements

A specification statement is a list of changing variables called the *frame* (say w), a formula called the *precondition* (say *pre*), and a formula called the *postcondition* (say *post*). Together they are written

w: [*pre* , *post*].

Informally this construct denotes a program which,

> if *pre* is true in the initial state, will establish *post* in the final state by changing only variables mentioned in the list w.

For the precise meaning, we have

Definition 4 *Specification statement:* For formulae *pre*, *post* over the program variables, and list of variables w,

$$[\![w\colon [pre\ ,\ post]\ ]\!]\psi \,\widehat{=}\, pre \wedge (\forall\, w\ \bullet\ post \Rightarrow \psi)$$

♡

The symbol $\widehat{=}$ is read "is defined to be equal to".

Specification statements allow program development to proceed at the level of refinement steps $\sqsubseteq$ rather than directly in terms of weakest preconditions, and are discussed in detail in [44, 48]. They are similar to the *descriptions* of [5] and the *prescriptions* of [50]. For now we extract from the above works a collection of *refinement laws*, given in the appendix to this paper. We illustrate their use with the following small program development:

> "assign to y the absolute value of x"
> $= y\colon [\mathit{true}\ ,\ y = |x|\,]$

$$= y: [(x \leq 0) \vee (x \geq 0) , y = |x|\,]$$

$\sqsubseteq$ Law 13
$$\textbf{if } x \leq 0 \to y: [x \leq 0 , y = |x|\,]$$
$$[\!]\; x \geq 0 \to y: [x \geq 0 , y = |x|\,]$$
$$\textbf{fi}$$

$$= \textbf{if } x \leq 0 \to y: [-x = |x| , y = |x|\,]$$
$$[\!]\; x \geq 0 \to y: [x = |x| , y = |x|\,]$$
$$\textbf{fi}$$

$\sqsubseteq$ Law 12 twice
$$\textbf{if } x \leq 0 \to y := -x$$
$$[\!]\; x \geq 0 \to y := x$$
$$\textbf{fi}$$

3.2 Program conjunction

Given a program P we write the generalised *program conjunction* of P over some variable i as $|[\,\textbf{con}\; i \bullet P\,]|$. We call it conjunction because that new program is a refinement $\sqsubseteq$ of the original program P for *all* values of the *logical constant* i. For example, consider the statement $x: [x = i , x = i + 1]$, and suppose our variables range over the natural numbers. Its generalised conjunction over i refines all of the following:

$$x: [x = 0 , x = 1]$$
$$x: [x = 1 , x = 2]$$
$$x: [x = 2 , x = 3]$$
$$\vdots$$

Each of those programs deals with a specific value of x, and can abort for all others. Yet, as Definition 5 will show, that generalised conjunction equals the statement $x := x + 1$, which is guaranteed to terminate.

Definition 5 *Program conjunction:* For program P and variable i not free in ψ,

$$[\![\,|[\,\textbf{con}\; i \bullet P\,]|\,]\!]\psi \mathrel{\widehat{=}} (\exists i \bullet [\![P]\!]\psi)$$

♡

As in Definition 2, systematic renaming can deal with occurrences of i in ψ.

Thus for the example above we can calculate

$$
\begin{aligned}
&\ [\![\,|[\,\mathbf{con}\ i \bullet i\colon [x = i\,,\ x = i+1]\,]|\,]\!]\psi \\
=&\ (\exists i \bullet [\![\,i\colon [x = i\,,\ x = i+1]\,]\!]\psi) & \text{Definition 5}\\
=&\ (\exists i \bullet x = i \wedge (\forall x \bullet x = i+1 \Rightarrow \psi)) & \text{Definition 4}\\
=&\ (\exists i \bullet x = i \wedge \psi[x\backslash i+1]) \\
=&\ \psi[x\backslash i+1][i\backslash x] \\
=&\ \psi[x\backslash x+1] & i \text{ not free in } \psi\\
=&\ [\![x := x+1]\!]\psi
\end{aligned}
$$

The notation $[x\backslash i+1]$ indicates syntactic replacement of x by $i+1$ with any changes of bound variable necessary to avoid capture.

Variables declared by **con** we call *logical constants*. They usually appear in program developments where some initial value must be fixed, in order to allow later reference to it. For example in the Hoare style [28], we might write "find a program P, changing only x, such that $\{x = X\}P\{x = X + 1\}$". Here the upper case X makes use of a convention that such variables are not to appear in the final program: it is not $x := X + 1$ that is sought, but $x := x + 1$. We would just write

$$|[\,\mathbf{con}\ X \bullet x\colon [x = X\,,\ x = X+1]\,]|,$$

$\hookleftarrow$ it being understood that we are looking for a refinement of that. Since our final programming language does not allow declarations **con**, we are forced to use refinements whose effect is to eliminate X. We do not need an upper-case convention.

It is interesting that program conjunction is the dual of local variable declaration (compare Definitions 2 and 5); thus logical constants are in that sense dual to local variables. It is shown in [20] that data refinement distributes through program conjunction.

4 Data refinement calculators

In Section 2 we defined the relation $\preceq$ of data-refinement between two statements S and S'. We gave there also a sufficient relation between the abstract initialisation I and the concrete initialisation I'.

In this section we show how the extensions of Section 3 allow us to *calculate* data-refinements S' and I' which satisfy the sufficient relations automatically. Following [35], we call these techniques *calculators*.

For the rest of this section, we will assume that the data-refinement is given by

abstract variables: a
concrete variables: c
abstraction invariant: AI

Moreover, we assume that the concrete variables c do not appear free in the abstract program.

4.1 The *initialisation* calculator

For concrete initialisation I' to data-refine the abstract I we know from Theorem 1 that $I' \Rightarrow (\exists a \bullet AI \wedge I)$ is sufficient; therefore we define I' to be $(\exists a \bullet AI \wedge I)$ itself. Law 5 (appendix) shows that we lose no generality, since any concrete initialisation I', where $I' \Rightarrow (\exists a \bullet AI \wedge I)$, can be reached in two stages: first replace I by the calculated $(\exists a \bullet AI \wedge I)$; then strengthen that, by Law 5, to I'.

If AI is *false*, then the calculated I' will be *false* also; indeed, Law 5 allows a refinement step to *false* initialisation directly. That is valid, though impractical, for the following reason: Definition 2 shows that the resulting program is *miraculous*:

$$[\![\, |[\, \textbf{var} \; l \mid false \bullet P \,]| \,]\!] false = true.$$

It can never be implemented in a programming language. (And that is why programming languages do not have empty types.)

4.2 The *specification* calculator

Lemma 3 to follow gives us a calculator for the data-refinement of any abstract statement of the form $a, x \colon [pre \, , \, post]$, where a and x are disjoint (and either may be empty). Lemma 4 shows that taking that data-refinement loses no generality. The two results are combined in Theorem 2. Finally, we give as a corollary a calculator for statements $b, x \colon [pre \, , \, post]$ where b is a subset of a; that is an abstract statement which may require some abstract variables not to change.

Lemma 3 *Validity:* The following data-refinement is always valid:

$$a, x \colon [pre \, , \, post]$$
$$\preceq \; c, x \colon [(\exists a \bullet AI \wedge pre) \, , \, (\exists a \bullet AI \wedge post)]$$

Proof: We take any formula ψ containing no free c, and proceed as follows:

$$
\begin{aligned}
&\quad (\exists\, a \bullet AI \wedge [\![\, a, x\colon [pre\,,\ post]\,]\!]\psi) \\
&= (\exists\, a \bullet AI \wedge pre \wedge (\forall\, a, x \bullet post \Rightarrow \psi)) && \text{Definition 4} \\
&= (\exists\, a \bullet AI \wedge pre) \wedge (\forall\, c, a, x \bullet post \Rightarrow \psi) && c \text{ not free in } post,\ \psi \\
&\Rightarrow (\exists\, a \bullet AI \wedge pre) \wedge (\forall\, c, x, a \bullet AI \wedge post \Rightarrow AI \wedge \psi) \\
&\Rightarrow (\exists\, a \bullet AI \wedge pre) \wedge (\forall\, c, x \bullet (\exists\, a \bullet AI \wedge post) \Rightarrow (\exists\, a \bullet AI \wedge \psi)) \\
&= [\![\, c, x\colon [(\exists\, a \bullet AI \wedge pre)\,,\ (\exists\, a \bullet AI \wedge post)]\,]\!]\, (\exists\, a \bullet AI \wedge \psi)
\end{aligned}
$$

♡

Lemma 4 *Generality:* For all programs CP, if $a, x\colon [pre\,,\ post] \preceq CP$ then

$$c, x\colon [(\exists\, a \bullet AI \wedge pre)\,,\ (\exists\, a \bullet AI \wedge post)] \sqsubseteq CP$$

Proof: We take any ψ containing no free a, and proceed as follows:

$$
\begin{aligned}
&\quad [\![\, c, x\colon [(\exists\, a \bullet AI \wedge pre)\,,\ (\exists\, a \bullet AI \wedge post)]\,]\!]\psi \\
&= (\exists\, a \bullet AI \wedge pre) \\
&\quad \wedge (\forall\, c, x \bullet (\exists\, a \bullet AI \wedge post) \Rightarrow \psi) && \text{Definition 4} \\
&= (\exists\, a \bullet AI \wedge pre) && c \text{ not free in } post, \\
&\quad \wedge (\forall\, a, x \bullet post \Rightarrow (\forall\, c \bullet AI \Rightarrow \psi)) && a \text{ not free in } \psi \\
&= (\exists\, a \bullet AI \wedge pre \wedge (\forall\, a, x \bullet post \Rightarrow (\forall\, c \bullet AI \Rightarrow \psi))) \\
&= (\exists\, a \bullet AI \wedge [\![a, x\colon [pre\,,\ post]]\!]\, (\forall\, c \bullet AI \Rightarrow \psi)) && \text{Definition 4} \\
&\Rightarrow [\![CP]\!]\,(\exists\, a \bullet AI \wedge (\forall\, c \bullet AI \Rightarrow \psi)) && \text{assumption, Definition 3} \\
&\Rightarrow [\![CP]\!]\psi && a \text{ not free in } \psi,\text{ monotonicity}
\end{aligned}
$$

♡

We now have the specification calculator we require: Lemma 3 states that it *is* a data refinement; Lemma 4 states that any other data-refinement of the abstract specification is an *algorithmic* refinement of the calculated one. We summarise that in Theorem 2:

Theorem 2 *The specification calculator:* For all programs CP,

$$a, x\colon [pre\,,\ post] \preceq CP$$

if and only if

$$c, x\colon [(\exists\, a \bullet AI \wedge pre)\,,\ (\exists\, a \bullet AI \wedge post)] \sqsubseteq CP$$

Proof: From Lemmas 3 and 4.
♡

Note that the quantifications ($\exists a \cdots$) ensure that the abstract variables a do not appear in the concrete program.

We conclude this section with a corollary of Lemma 3; it calculates the data-refinement of an abstract specification in which not all variables are changing. In its proof we are able to reason at the higher level of the relations $\sqsubseteq$ and $\preceq$; weakest preconditions are not required.

This corollary is the first occasion we have to use logical constants in data refinement. Like local variables, logical constants are *bound* in a program; and it is the **con** declaration which binds the abstract variables a in Corollary 1, since the quantification ($\exists b \cdots$) alone may leave some abstract variables free.

Corollary 1 For any subset (not necessarily proper) b of the abstract variables a, the abstract specification b, x: $[pre \ , \ post]$ is data-refined by

$$|[\ \textbf{con } a \ \bullet$$
$$\quad c, x \colon [AI \wedge pre \ , \ (\exists b \ \bullet \ AI \wedge post)]$$
$$]|$$

Proof: Let b and y partition a, and let B and Y partition A correspondingly. Then

$\quad b, x \colon [pre \ , \ post]$

$= $ Law 9
$\quad |[\ \textbf{con } Y \ \bullet \ b, y, x \colon [pre \wedge y = Y \ , \ post \wedge y = Y] \ ]|$

$\preceq$ Lemma 3
$\quad |[\ \textbf{con } Y \ \bullet$
$\quad\quad c, x \colon [(\exists b, y \ \bullet \ AI \wedge pre \wedge y = Y) \ , \ (\exists b, y \ \bullet \ AI \wedge post \wedge y = Y)]$
$\quad]|$

$= \ |[\ \textbf{con } Y \ \bullet$
$\quad\quad c, x \colon [(\exists b \ \bullet \ AI \wedge pre)[y \backslash Y] \ , \ (\exists b \ \bullet \ AI \wedge post)[y \backslash Y]]$
$\quad]|$

$= $ Law 8
$\quad |[\ \textbf{con } y \ \bullet$
$\quad\quad c, x \colon [(\exists b \ \bullet \ AI \wedge pre) \ , \ (\exists b \ \bullet \ AI \wedge post)]$
$\quad]|$

$= $ Law 6
$\quad |[\ \textbf{con } a \ \bullet$
$\quad\quad c, x \colon [AI \wedge pre \ , \ (\exists b \ \bullet \ AI \wedge post)]$
$\quad]|$

♡

4.3 The *guard* calculator

We saw in Corollary 1 that the specification calculator introduces **con** a and existentially quantifies over changing abstract variables only. For guards, changing nothing, we would expect that quantification to be empty. We have

Theorem 3 *The guard calculator:* If $S_i \preceq S'_i$ for each i, then the following refinement is valid:

$$\begin{array}{l} \textbf{if } (\![i \bullet G_i \to S_i) \textbf{ fi} \\ \preceq |[\, \textbf{con } a \bullet \\ \quad \textbf{if } (\![i \bullet AI \wedge G_i \to S'_i) \textbf{ fi} \\ \,]| \end{array}$$

Proof: For any ψ not containing c, we have

$$\begin{array}{rl} & (\exists a \bullet AI \wedge [\![\textbf{if } (\![i \bullet G_i \to S_i) \textbf{ fi}]\!] \psi) \\ = & \qquad\qquad\qquad\qquad\qquad\qquad\qquad \text{definition } [\![\textbf{if} \cdots \textbf{fi}]\!] \\ & (\exists a \bullet AI \wedge (\bigvee i \bullet G_i) \wedge (\bigwedge i \bullet G_i \Rightarrow [\![S_i]\!] \psi)) \\ = & (\exists a \bullet (\bigvee i \bullet AI \wedge G_i) \wedge (\bigwedge i \bullet AI \wedge G_i \Rightarrow AI \wedge [\![S_i]\!] \psi)) \\ \Rightarrow & (\exists a \bullet (\bigvee i \bullet AI \wedge G_i) \wedge (\bigwedge i \bullet AI \wedge G_i \Rightarrow (\exists a \bullet AI \wedge [\![S_i]\!] \psi))) \\ \Rightarrow & \qquad\qquad\qquad\qquad\qquad\qquad\qquad \text{since } S_i \preceq S'_i \\ & (\exists a \bullet (\bigvee i \bullet AI \wedge G_i) \wedge (\bigwedge i \bullet AI \wedge G_i \Rightarrow [\![S'_i]\!] (\exists a \bullet AI \wedge \psi))) \\ = & [\![\, |[\textbf{con } a \bullet \cdots]| \,]\!] (\exists a \bullet AI \wedge \psi) \end{array}$$

♡

A similar construction is possible for **do** $\cdots$ **od**, but in this general setting it is better to use **if** $\cdots$ **fi** and recursion. There are special cases for **do**, however, and they are discussed in Section 6.

5 Example of refinement: the "mean" module

We can present a data refinement independently of its surrounding program text by collecting together all the statements that refer to the abstract variables or to variables in the abstraction invariant. Such a collection is called a *module*, and we can confine our attention to it for this reason: statements which do

```
module Calculator ≙
  var b: bag of Real;

  procedure Clear ≙ b := ≺≻ ;
  procedure Enter (value r) ≙ b := b+ ≺ r ≻ ;
  procedure Mean (result m) ≙
    if b ≠ ≺≻ → m := ∑ b/#b
    ▯ b = ≺≻ → error
    fi
end
```

Figure 1: The "mean" module

not refer to abstract variables, or to the abstraction invariant, are refined by themselves and we need not change them.

Consider the module of Figure 1 for calculating the mean of a sample of numbers. We write bag comprehensions between brackets ≺≻, and use $\sum b$ and $\#b$ for the sum and size respectively of bag b. The operator $+$ is used for bag addition. The statement **error** is some definite error indication, and we assume that **error** $\preceq$ **error**. The initialisation is $b \in$ **bag of** *Real*.

The module is operated by: first *clearing*; then *entering* the sample values, one at a time; then finally taking the *mean* of all those values.

For the data refinement, we represent the bag by its sum s and size n at any time.

> abstract variables: b
> concrete variables: s, n
> abstraction invariant: $s = \sum b \land n = \#b$

We data-refine the module by replacing the abstract variables b by the concrete variables s, n and applying the calculations of Section 4 to the initialisation and the three procedures. Stacked formulae below denote their conjunction.

- For the initialisation, we have from Section 4.1 for the concrete initialisation

$$\left(\exists b \bullet \begin{array}{c} b \in \textbf{bag of } Real \\ s = \sum b \\ n = \#b \end{array} \right)$$

$$= \left(\begin{array}{c} s \in Real \\ n \in Natural \\ n = 0 \Rightarrow s = 0 \end{array} \right)$$

- For the procedure *Clear*, we have from Section 4.2

    $b := \prec\succ$
    $= b: [true, b = \prec\succ]$
    $\preceq$ Lemma 3
    $$s, n: \left[\left(\exists b \bullet \begin{matrix} s = \sum b \\ n = \#b \end{matrix} \right), \left(\exists b \bullet \begin{matrix} s = \sum b \\ n = \#b \\ b = \prec\succ \end{matrix} \right) \right]$$
    $\sqsubseteq$ Law 1
    $s, n: [true, s = 0 \wedge n = 0]$
    $\sqsubseteq$ Law 12
    $s, n := 0, 0$

- For the procedure *Enter*, we have from Section 4.2

    $b := b + \prec r \succ$
    $= |[\text{con } B \bullet b: [b = B, b = B + \prec r \succ]]|$
    $\preceq$ Lemma 3
    $|[\text{con } B \bullet s, n: \begin{bmatrix} s = \sum B & s = \sum(B + \prec r \succ) \\ n = \#B & , n = \#(B + \prec r \succ) \end{bmatrix}]|$
    $\sqsubseteq$ Laws 12, 7
    $s, n := s + r, n + 1$

- For *Mean* we have first that from Section 4.2

    $m := \sum b / \#b$
    $= m: [\#b \neq 0, m = \sum b / \#b]$
    $\preceq$ Corollary 1 (noting the quantification is empty)
    $|[\text{con } b \bullet$
    $$m, s, n: \begin{bmatrix} \#b \neq 0 & m = \sum b/\#b \\ s = \sum b & , s = \sum b \\ n = \#b & n = \#b \end{bmatrix}$$
    $]|$
    $\sqsubseteq$ Laws 10, 2, 3, 1
    $|[\text{con } b \bullet m: [n \neq 0, m = s/n]]|$
    $\sqsubseteq$ Laws 12, 7
    $m := s/n$

Then we conclude from Theorem 3 that

if $b \neq \prec\succ \rightarrow m := \sum b / \#b$
[] $b = \prec\succ \rightarrow$ **error**
fi

$$\preceq \; |[\; \mathbf{con} \; b \; \bullet$$
$$\mathbf{if} \begin{pmatrix} b \neq \prec \succ \\ s = \sum b \\ n = \# b \end{pmatrix} \to m := s/n$$
$$[] \begin{pmatrix} b = \prec \succ \\ s = \sum b \\ n = \# b \end{pmatrix} \to \mathbf{error}$$
$$\mathbf{fi}$$
$$]|$$

To make further progress with *Mean*, we need to eliminate the abstract variable b from the guards; then Law 7 applies. That is assisted by the following lemma (which is generally applicable to the refinement of alternations, whether or not they occur within data refinements):

Lemma 5 *Refining guards:* Given the conditions

1. $(\bigvee i \bullet G_i) \Rightarrow (\bigvee i \bullet G'_i)$
2. $(\bigvee i \bullet G_i) \Rightarrow (G'_i \Rightarrow G_i)$ *for each* i

the following refinement is valid:

$$\mathbf{if} \; ([] i \bullet G_i \to S_i) \; \mathbf{fi} \; \sqsubseteq \; \mathbf{if} \; ([] i \bullet G'_i \to S_i) \; \mathbf{fi}$$

Proof: By Lemma 1 and $[\![\mathbf{if} \cdots \mathbf{fi}]\!]$ we must show for all formulae ψ that

$$(\bigvee i \bullet G_i) \wedge (\bigwedge i \bullet G_i \Rightarrow [\![S_i]\!]\psi)$$
$$\Rightarrow (\bigvee i \bullet G'_i) \wedge (\bigwedge i \bullet G'_i \Rightarrow [\![S_i]\!]\psi)$$

That follows by propositional calculus from assumptions 1 and 2 above.
♡

We have immediately the following corollary:

Corollary 2 *Weakening guards*: The following refinement is valid for *any* formula X:

$$\mathbf{if} \; ([] i \bullet G_i \wedge X \to S_i) \; \mathbf{fi} \; \sqsubseteq \; \mathbf{if} \; ([] i \bullet G_i \to S_i) \; \mathbf{fi}$$

♡

module *Calculator* $\stackrel{\frown}{=}$
  var s: *Real*, n: *Natural*;

  procedure *Clear* $\stackrel{\frown}{=} s, n := 0, 0$;
  procedure *Enter*(**value** r) $\stackrel{\frown}{=} s, n := s + n, n + 1$;
  procedure *Mean*(**result** m) $\stackrel{\frown}{=}$
    if $n \neq 0 \rightarrow m := s/n$
    [] $n = 0 \rightarrow$ **error**
    fi

  initially $n = 0 \Rightarrow s = 0$
end

Figure 2: The "mean" module, after data refinement

Now we can continue the refinement of *Mean*:

$\sqsubseteq$ Lemma 5, Law 7
  if $n \neq 0 \rightarrow m := s/n$
  [] $n = 0 \rightarrow$ **error**
  fi

In Figure 2 we give the resulting data refinement for the whole module.

To see the need for the initialisation, consider this alternative definition of *Clear*:

procedure *Clear* $\stackrel{\frown}{=}$
  if $b \neq \prec\succ \rightarrow b := \prec\succ$
  [] $b = \prec\succ \rightarrow$ **skip**
  fi

That is semantically identical to the original, in Figure 1, but might be cheaper overall if the operation $b := \prec\succ$ were expensive. Its calculated data refinement is

procedure *Clear* $\stackrel{\frown}{=}$
  if $n \neq 0 \rightarrow s, n := 0, 0$
  [] $n = 0 \rightarrow$ **skip**
  fi

That would *not* work correctly if used immediately after an initialisation, say, of $s = 1 \land n = 0$! So our stated initialisation is necessary, after all; note however that since initialisations can always be strengthened (Law 5), we could use the simpler $s = 0$ if desired.

6 Specialised techniques

Now we specialise the techniques of Section 4: we consider *guards*, *functional* data-refinement, and the use of *auxiliary variables*.

6.1 Data-refining guards

We have seen that data refinement takes an abstract guard G to a concrete guard $G \wedge AI$, where AI is the abstraction invariant. The occurrences of abstract variables in this concrete guard must then be eliminated. We use Lemma 5 for that: we replace each of the calculated guards $G_i \wedge AI$ by the guard $(\forall a \bullet AI \Rightarrow G_i)$, which does not contain a free. By that lemma, we must show

1. $(\bigvee i \bullet G_i \wedge AI) \Rightarrow (\bigvee i \bullet (\forall a \bullet AI \Rightarrow G_i))$
2. $(\bigvee i \bullet G_i \wedge AI) \Rightarrow ((\forall a \bullet AI \Rightarrow G_i) \Rightarrow G_i \wedge AI)$ *for each i*

The validity of 2 is evident; and by rewriting 1 we can see that it requires only that the data-refined *disjunction* of the abstract guards implies the disjunction of the concrete guards. Thus we have the following

Lemma 6 *Data refinement of alternations:* Given abstraction invariant AI, abstract guards G_i, and abstract statements S_i, let the concrete guards G'_i and concrete statements S'_i be such that

1. $G'_i = (\forall a \bullet AI \Rightarrow G_i)$
2. $S_i \preceq S'_i$

Then provided $(\exists a \bullet AI \wedge (\bigvee i \bullet G_i)) \Rightarrow (\bigvee i \bullet G'_i)$, the following data refinement is valid:

$$\text{if } ([\!]\, i \bullet G_i \to S_i) \text{ fi } \preceq \text{ if } ([\!]\, i \bullet G'_i \to S'_i) \text{ fi}$$

♡

For iterations the result is the same: we use the recursive formulation

$$\text{do } ([\!]\, i \bullet G_i \to S_i) \text{ od } \stackrel{\wedge}{=} (\mu P \bullet \text{if} \quad ([\!]\, i \bullet G_i \to S_i;\ P) \\ [\!] \quad \neg(\bigvee i \bullet G_i) \to \text{skip} \\ \text{fi})$$

and hence must determine the conditions under which

$$\begin{aligned}&\textbf{if } (\lbrack\!\rbrack i \bullet AI \wedge G_i \to S_i;\ P) \\ &\lbrack\!\rbrack\ AI \wedge \neg(\bigvee i \bullet G_i) \to \textbf{skip} \\ &\textbf{fi}\end{aligned}$$

$$\sqsubseteq \begin{aligned}&\textbf{if } (\lbrack\!\rbrack i \bullet G'_i \to S'_i;\ P) \\ &\lbrack\!\rbrack\ \neg(\bigvee i \bullet G'_i) \to \textbf{skip} \\ &\textbf{fi}\end{aligned}$$

As before we have defined G'_i to be $(\forall a \bullet AI \Rightarrow G_i)$. Straightforward application of Lemma 5 gives us

Lemma 7 *Data refinement of iterations:* Under the same conditions as Lemma 6, the following refinement is valid:

$$\textbf{do } (\lbrack\!\rbrack i \bullet G_i \to S_i)\ \textbf{od}\ \preceq\ \textbf{do } (\lbrack\!\rbrack i \bullet G'_i \to S'_i)\ \textbf{od}$$

♡

Our choice of G'_i is used also in [51], where those two rules are proved from first principles (that is, from Definition 3). We have shown therefore how that technique is an instance of our Theorem 3.

6.2 Functional refinement

In many cases, the abstraction invariant is *functional* in the sense that for any concrete value there is at most one corresponding abstract value. In [33], for example, this is the primary form of data-refinement considered.

Functional abstraction invariants can always be written as a conjunction

$a = AF(c)$
$CI(c)$

where AF we call the *abstraction function* and CI the *concrete invariant*; the formula CI of course contains no occurrences of abstract variables a. We assume that $CI(c)$ implies well-definedness of AF at c.

Functional data-refinements usually lead to simpler calculations. First, the concrete formula $(\exists a \bullet AI \wedge \phi)$ — where ϕ is *pre* or *post* in the abstract specification — is simplified:

$$\begin{aligned}&(\exists a \bullet AI \wedge \phi) \\ =\ &(\exists a \bullet (a = AF(c)) \wedge CI(c) \wedge \phi) \\ =\ &CI(c) \wedge \phi[a \backslash AF(c)]\end{aligned}$$

Thus in this case data-refinement calculations are no more than simple substitutions. Note also that the resulting concrete formula contains no free abstract variables, and this allows any $|[$ **con** $a \bullet \cdots]|$ to be eliminated immediately. We have this corollary of Theorem 2:

Corollary 3 *Functional data-refinement*: Given an abstraction invariant $a = AF(c) \wedge CI(c)$, the following data-refinement is always valid:

$$a, x: \lfloor pre \; , \; post \rfloor$$

$$\preceq \; c, x: \begin{bmatrix} pre[a \backslash AF(c)] & , & post[a \backslash AF(c)] \\ CI(c) & & CI(c) \end{bmatrix}$$

Moreover, it is the most general.
♡

A second advantage is in the treatment of guards, as is shown also in [51]. We replace as before G_i by $G_i \wedge AI$, which becomes

$$G_i \wedge (a = AF(c)) \wedge CI(c)$$
$$= G_i[a \backslash AF(c)] \wedge (a = AF(c)) \wedge CI(c)$$

Now by Corollary 2, we can eliminate the conjunct $a = AF(c)$ immediately, and hence the enclosing $|[$ **con** $a \bullet \cdots]|$ as well. (And we can eliminate the $CI(c)$, but that is optional: it contains no a.) So we have the following result for the functional data-refinement of alternations:

Lemma 8 *Functional data-refinement of alternations:* Given abstraction invariant $(a = AF(c)) \wedge CI(c)$, abstract guards G_i, and abstract statements S_i, let concrete guards G'_i and concrete statements S'_i be such that

1. $G'_i = G_i[a \backslash AF(c)] \wedge CI(c)$
2. $S_i \preceq S'_i$

Then the following data refinement is always valid

if $([]i \bullet G_i \to S_i)$ **fi** $\preceq$ **if** $([]i \bullet G'_i \to S'_i)$ **fi**

♡

The same remarks apply to iteration (and again, the conjunct $CI(c)$ is optional in the concrete guards):

Lemma 9 *Functional data-refinement of iterations:* Under the same conditions as Lemma 8, the following data refinement is valid

$$\mathbf{do}\ (\![i \bullet G_i \rightarrow S_i)\ \mathbf{od}\ \preceq\ \mathbf{do}\ (\![i \bullet G_i' \rightarrow S_i')\ \mathbf{od}$$

♡

6.3 Auxiliary variables

A set of local variables is *auxiliary* if its members occur only in statements which assign to members of that set. They can be used for data refinement as follows.

There are three stages. In the first, an abstraction invariant is chosen, relating abstract variables to concrete. Declarations of those concrete variables are added to the program, but the declarations of the abstract variables are *not* removed. The initialisation is strengthened so that it implies the abstraction invariant; every guard is strengthened by conjoining the abstraction invariant; and every assignment statement is extended, if necessary, by assignments to concrete variables which *maintain* the the abstraction invariant.

In the second stage, the program is *algorithmically* refined so that the abstract variables become auxiliary. In the third stage, the (now) auxiliary abstract variables are removed (their declarations too), leaving only the concrete — and the data-refinement is complete.

That technique was proposed by [37], and a simple example is given in [17, p.64]. It is a special case of our present technique, as we now show. Suppose our overall aim is the following data-refinement:

      abstract variables: a
      concrete variables: c
      abstraction invariant: AI

We decompose this into two data-refinements, applied in succession. In the first, there are *no* abstract variables:

      abstract variables: (none)
      concrete variables: c
      abstraction invariant: AI

Clearly this removes no declarations, and from Definition 3 requires for $S \preceq S'$ (remembering that the quantification $(\exists\, a \bullet \cdots)$ is empty) only that for all ψ not containing c free, we have

$$AI \wedge [\![S]\!]\psi \Rightarrow [\![S]\!](AI \wedge \psi)$$

That is precisely the first stage explained informally above.

The second stage remains: it is only algorithmic refinement. For the third stage, we use the following data refinement in which there are no *concrete* variables:

abstract variables:	a
concrete variables:	(none)
abstraction invariant:	*true*

From Definition 3, here for $S \preceq S'$ we must show that for all formulae ψ

$$(\exists a \bullet [\![S]\!]\psi) \Rightarrow [\![S']\!](\exists a \bullet \psi)$$

And this holds only when the abstract variables a are auxiliary.

We illustrate the auxiliary technique with two lemmas, derived from our general rules for data refinement:

Lemma 10 *Introducing concrete variables while maintaining the invariant:* Let the abstract variables be *none*, the concrete variables be c, and the abstraction invariant AI. Then for abstract expression AE and concrete expression CE, we have

$$a := AE \preceq a, c := AE, CE$$

provided $AI \Rightarrow [\![a, c := AE, CE]\!]AI$.

Proof:

$$
\begin{aligned}
& AI \wedge [\![a := AE]\!]\psi \\
&= AI \wedge \psi[a\backslash AE] && \text{by semantics of } := \\
&\Rightarrow [\![a, c := AE, CE]\!]AI \wedge \psi[a\backslash AE] && \text{by assumption} \\
&\Rightarrow AI[a, c\backslash AE, CE] \wedge \psi[a\backslash AE] && \text{by semantics of } := \\
&= AI[a, c\backslash AE, CE] \wedge \psi[a, c\backslash AE, CE] && \text{since } \psi \text{ contains no } c \\
&= [\![a, c := AE, CE]\!](AI \wedge \psi) && \text{by semantics of } :=
\end{aligned}
$$

♡

Lemma 11 *Eliminating auxiliary variables:* Let the abstract variables be a, the concrete variables be *none*, and the abstraction invariant *true*. Then

1. $a := AE \preceq \textbf{skip}$

2. $c := CE \preceq c := CE$

provided CE contains no occurrence of a.

Proof: For 1 we have

$(\exists a \bullet [\![a := AE]\!]\psi)$
$= (\exists a \bullet \psi[a\backslash AE])$ by semantics of :=
$\Rightarrow (\exists a \bullet \psi)$ predicate calculus
$= [\![\mathbf{skip}]\!](\exists a \bullet \psi)$

For 2 we have

$(\exists a \bullet [\![c := CE]\!]\psi)$
$= (\exists a \bullet \psi[c\backslash CE])$ by semantics of :=
$= (\exists a \bullet \psi)[c\backslash CE]$ since CE contains no a
$= [\![c := CE]\!](\exists a \bullet \psi)$

(Note that in case 2 we did not assume that ψ contained no c.)

♡

If the abstract statement is a specification $a: [pre , post]$, then in the first stage we replace it by $a, c: [pre \wedge AI , post \wedge AI]$. If by the third stage (after algorithmic refinement) we still have a specification — say $a, c: [pre' , post']$, then the removal of a as an auxiliary variable leaves us with the specification $c: [(\exists a \bullet pre') , (\exists a \bullet post')]$.

Let us as a final illustration try to remove a variable which is *not* auxiliary: we take the data-refinement as for the third stage, and suppose that $c := a \preceq CP$ for some concrete program CP. We expect this to fail, since a is clearly not auxiliary in $c := a$. Now we have for all constants n that

$true$
$= (\exists c \bullet c = n)$ predicate calculus
$= (\exists a \bullet (c = n)[c\backslash a])$ renaming bound variable c to a
$= (\exists a \bullet [\![c := a]\!](c = n))$ by semantics of :=
$\Rightarrow [\![CP]\!](\exists a \bullet c = n)$ by assumption
$= [\![CP]\!](c = n)$

Since the above holds for any n, we have that CP always establishes both $c = 0$ and $c = 1$. Because no executable program can do this, we have shown that there is no such CP — as hoped, a cannot be eliminated from $c := a$. But what if we write $c := a$ as a specification? In that case, we have

$c := a$

$= c\colon [\mathit{true}\ ,\ c = a]$

$\preceq$ Corollary 1 (noting the quantification is empty)
 |[**con** $a \bullet c\colon [\mathit{true}\ ,\ c = a]$]|

So here we *have* a data-refinement, after all. But that is consistent with the above in the following way: there is no executable program CP (whether containing a or not) such that $c\colon [\mathit{true}\ ,\ c = a] \sqsubseteq CP$. Thus the |[**con** $a \bullet \cdots$]| still cannot be eliminated.

In [41] the auxiliary variable technique is presented independently of the refinement calculus.

7 Conclusions

Our calculators for data refinement make it possible in principle to see that activity as the routine application of laws. The example of Section 5 is a demonstration for a simple case. It is important in practice, however, to take advantage of the specialised techniques of Section 6; otherwise, the subsequent algorithmic refinement will simply repeat the derivation of the techniques themselves, again and again.

That subsequent algorithmic refinement is in fact a lingering problem. In many cases, particularly with larger and more sophisticated refinements, the refined operations present fearsome collections of formulae concerning data structures for which we do not have an adequate body of theory. Their subsequent manipulations in the predicate calculus resemble programming in machine code. Fortunately, there is work on such theories (and *their* calculi, for example [14]), and we see little difficulty in taking advantage of them.

Our work on data refinement has been aided and improved by collaboration with Morris and Back, who present their work in [51] and [5] respectively. We extend Morris's approach by our use of logical constants . A second extension is our "if and only if" result in Theorem 2. That is necessary, we feel, for a data refinement to be called a calculator: $P \leq Q$ is a *calculator* only if taking Q loses no generality. And Morris retains some restrictions on abstraction invariants which we believe are unnecessary. Conversely, Morris's specialised alternation calculator [51, Theorem 4] improves ours (Lemma 6) by introducing a miracle as the refined program [42]; his rule needs no proof obligation. Our work extends Back's by our emphasis on calculation, and our use of logical constants.

8 Acknowledgements

We are grateful to have had the opportunity to discuss our work with Ralph Back and Joe Morris, and for the comments made by members of IFIP WG 2.3. Much of our contact with other researchers has been made possible by the generosity of British Petroleum Ltd.

9 Appendix: refinement laws

Below is a collection of laws which can in principle take most specification statements through a series of refinements into executable code. We have not tried to make them complete. "Executable code" means program text which does not include either specification statements or logical constants.

"In principle" means that these basic rules, used alone, will in many cases give refinement sequences which are very long indeed — rather like calculating derivatives from first principles. But with experience, one collects a repertoire of more powerful and specific laws which make those calculations routine.

Some of the laws below are equalities $=$; some are proper refinements $\sqsubseteq$. In all cases they have been proved using the *weakest precondition* semantics of the constructs concerned.

Section 9.2 contains notes relating to the laws of Section 9.1.

9.1 Laws of program refinement

Most of these laws are extracted from [48], retaining only those used in this paper. Logical constant laws have been added.

1. *Weakening the precondition:* If $pre \Rightarrow pre'$ then

    $w: [pre , post] \sqsubseteq w: [pre' , post]$

2. *Strengthening the postcondition:* If $post' \Rightarrow post$ then

    $w: [pre , post] \sqsubseteq w: [pre , post']$

3. *Assuming the precondition in the postcondition:*

    $w: [pre , (\exists w \bullet pre) \land post] = w: [pre , post]$

4. *Introducing local variables:* If x does not appear free in *pre* or *post*, then

    $w: [pre , post] \sqsubseteq |[\mathbf{var}\ x \mid I \bullet w,x: [pre , post]]|$

5. *Strengthening the initialisation:* If $I' \Rightarrow I$, then

$$|[\,\textbf{var}\ x\mid I\bullet S\,]| \sqsubseteq |[\,\textbf{var}\ x\mid I'\bullet S\,]|$$

6. *Introducing logical constants:* If x does not appear free in *post*, then

$$w\colon [(\exists x\bullet pre)\,,\ post] = |[\,\textbf{con}\ x\bullet w\colon [pre\,,\ post]\,]|$$

7. *Eliminating logical constants:* If x does not appear free in P, then

$$|[\,\textbf{con}\ x\bullet P\,]| = P$$

8. *Renaming logical constants:* If y is disjoint from w, and does not occur free in *pre* or *post*, then

$$|[\,\textbf{con}\ x\bullet w\colon [pre\,,\ post]\,]|$$
$$= |[\,\textbf{con}\ y\bullet w[x\backslash y]\colon [pre[x\backslash y]\,,\ post[x\backslash y]]\,]|$$

9. *Expanding the frame:* If x and y are fresh variables, disjoint from each other, then

$$w\colon [pre\,,\ post] = |[\,\textbf{con}\ y\bullet w,x\colon [pre\wedge x=y\,,\ post\wedge x=y]\,]|$$

10. *Contracting the frame:* If w and x are disjoint, then

$$w,x\colon [pre\,,\ post] \sqsubseteq w\colon [pre\,,\ post]$$

11. *Introducing* **skip**:

$$w\colon [post\,,\ post] \sqsubseteq \textbf{skip}$$

12. *Introducing assignment:* If E is an expression, then

$$w\colon [post[w\backslash E]\,,\ post] \sqsubseteq w:=E$$

13. *Introducing alternation:*

$$w\colon [pre\wedge(\bigvee i\bullet G_i)\,,\ post]$$
$$= \textbf{if}\,([\!]\,i\bullet G_i \to w\colon [pre\wedge G_i\,,\ post])\,\textbf{fi}$$

9.2 Notes

1. Law 3 applies when information from the precondition is needed in the postcondition. We use it below to derive a stronger version of Law 2:

   If $((\exists w\bullet pre)\wedge post') \Rightarrow post$, then
   $$w\colon [pre\,,\ post]$$

$\sqsubseteq$ by Law 2 and the assumption
   $w: [pre,\ (\exists\, w \bullet pre) \wedge post']$
   $\sqsubseteq$ by Law 3
   $w: [pre,\ post']$

2. Usually Law 6 is used to introduce an equality into the precondition which "saves an initial value for later." That is summarised in the following derived law:

   If y is disjoint from w, and does not occur free in *pre* or *post*, then

   $w: [pre,\ post]$
   $\sqsubseteq$ by Law 1
   $w: [(\exists\, y \bullet x = y \wedge pre),\ post]$
   $\sqsubseteq$ by Law 6
   $|[\mathbf{con}\ y \bullet w: [x = y \wedge pre,\ post]\,]|$

3. *Logical constants*, introduced by **con**, are variables which we can use during program development but not in final programs. Usually they are used to fix initial values, as in

   $|[\mathbf{con}\ X \bullet x: [x = X,\ x = X + 1]\,]|$
   $\sqsubseteq$ by Law 12
   $|[\mathbf{con}\ X \bullet x := x + 1\,]|$
   $\sqsubseteq$ by Law 7
   $x := x + 1$

   Since the keyword **con** is does not occur in our executable programming language — just as specification statements do not — it must be eliminated (using Law 7 as above) during the development process. Thus logical constants never appear in the final program, since they cannot be declared there.

4. Law 12 is usually applied together with Laws 10 and 1, as in the following derived rule:

   If the variables w and x are disjoint, E is an expression, and $pre \Rightarrow post[w\backslash E]$, then

   $w, x: [pre,\ post]$
   $\sqsubseteq$ by Law 10
   $w: [pre,\ post]$
   $\sqsubseteq$ by Law 1 and the assumption
   $w: [post[w\backslash E],\ post]$
   $\sqsubseteq$ by Law 12
   $w := E$

A Single Complete Rule for Data Refinement

P.H.B. Gardiner and Carroll Morgan

Abstract

One module is said to be refined by a second if no program using the second module can detect that it is not using the first; in that case the second module can replace the first in any program. Data refinement transforms the interior pieces of a module — its state and consequentially its operations — in order to refine the module overall.

A method for data refinement is sound if applying it actually does refine the module; a method is complete if any refinement of modules can be realised by its application.

It has been known for some time that there are two methods of data refinement which are jointly complete for boundedly-nondeterministic programs: any refinement can be realised by applying one method then the other. Those two methods are formulated in terms of relations between states. Here it is shown that using predicate transformers instead allows a single complete method.

1 Introduction

Data refinement transforms the interior of a module in order to effect a refinement overall of the module's external behaviour. Usually it is used, in program development, to replace abstract mathematical data structures by concrete structures that are more easily implemented [30, 33].

Hoare, He, and Sanders [32] give two methods of data refinement, which they call downward and upward simulation[1], and they show those methods to be sound and jointly complete. Abadi and Lamport [1] present similar but more comprehensive results that include stuttering and liveness. Both [32, 1] place their work in the context of state-to-state relations.

Those authors show by example that the earlier methods of [30, 33] are incomplete in this sense: there exist modules A and C, say, such that A is refined by C but it cannot be proved. Completeness is recovered by a new refinement method (upward simulation [32], or prophesy variables [1]), which

---

[1] We call these simulation and cosimulation respectively.

applies to some situations in which the usual method fails. Using both methods, in succession, any valid refinement can be shown — provided nondeterminism is bounded.

In [20] we used predicate transformers rather than relations to present a general method[2] of data refinement, and we showed relational downward simulation to be a special case of it. Morris [51] uses predicate transformers to present relational downward simulation specifically.

The contribution of this paper is our belated realisation that relational upward simulation, too, is a special case of our earlier work provided some of our constraints are relaxed. The result therefore is a single method complete on its own. Our first proof of this result [19] relied on the earlier work of Hoare, He, and Sanders [32]; here we give a more self-contained presentation, entirely in terms of predicate transformers, although we still borrow many of their ideas.

Sect. 2 introduces the important concepts relating to data refinement, including the definition of datatype and simulation, thus providing the context in which to reason about soundness and completeness. Sect. 3 introduces the properties of predicate transformers on which our results rely. In Sect. 4 and 5 the proofs of completeness and soundness are presented. At this stage, a restricted language is considered, in which only non-iterative programs are considered. In Sect. 6 the results of the previous two sections are generalised to include iteration and recursion. Lastly, in Sect. 7 we provide an example, demonstrating the use of our data-refinement rule.

2 Data Refinement

In this section we introduce definitions for the important concepts relating to data refinement. These definitions are effectively the same as those used by Hoare et. al. [26].

Within this section, "program" is considered as a primitive notion. One can think of programs as state-to-state relations or as predicate transformers, but we leave them uninterpreted for now, noting only the existence of $\sqsubseteq$ (a relation between programs) and a set of program-to-program functions, representing the operators of a programing language. We assume there is one associative binary operator $\fatsemi$.

For any particular interpretation, $\sqsubseteq$ will represent some form of refinement ordering, and $\fatsemi$ will represent some form of sequential composition. So $P \sqsubseteq Q$ can be read "P is refined by Q" (i.e., "anything P can do, Q can do better, or at least no worse"), and $P \fatsemi Q$ can be read "P then Q" (i.e., "the result of doing P and then doing Q").

A datatype is a set of programs for providing controlled access to local data;

---

[2] Actually just cosimulation again, but using predicate transformers instead of relations.

the set of programs are called the operations of the datatype. A program calling on a datatype has no other access to the local data, and the calling program's effect on the local data, via the datatype, is not considered as part of its external behaviour. This invisibility of local data can be made precise by stipulating that all executions of calling programs begin and end with distinguished operations, called the *initialisation* and *finalisation*. The initialisation creates and gives initial values to the local data, and the finalisation destroys them. These two operations are unusual in that their execution starts and finishes in different state spaces: the initialisation starts execution in the presence of the global variables only, and ends execution in the presence of global and local variables; the finalisation starts with global and local variables and ends with just global variables. In most programming languages the initialisation and finalisation are combined to form a single construct, which introduces and delimits the scope of local variables (i.e., a local-variable declaration block). Some languages use modules as another way to enforce this structure. In either case the finalisation is the projection function from the Cartesian product of global and local state spaces onto the global state space.

Definition 1 *A datatype is a triple* (I, OP, F), *where I and F are programs called respectively the initialisation and finalisation, and where OP is an indexed set of programs. Collectively, the members of OP together with I and F are called the operations of the datatype.*

♡

Note that Def. 1 does not mention the structure of state spaces; just the existence of the initialisation and finalisation suffices. The references to global and local data appear in the surrounding text only as an aid to the reader.

Refinement between programs is used to define the related notion of refinement between datatypes: one datatype is refined by another if any program that uses the former would function at least as well using the latter.

Definition 2 (I, OP, F) *is refined by* (I', OP', F')
if, for every function $\mathcal{P}$, expressible as a composition of program-language operators,

$$I \mathbin{\raisebox{0.5ex}{\scriptsize$\circ$}} \mathcal{P}(OP) \mathbin{\raisebox{0.5ex}{\scriptsize$\circ$}} F \sqsubseteq I' \mathbin{\raisebox{0.5ex}{\scriptsize$\circ$}} \mathcal{P}(OP') \mathbin{\raisebox{0.5ex}{\scriptsize$\circ$}} F'$$

♡

We usually refer to the datatype being refined as *abstract*, and to the other as *concrete*.

The above definition involves the use of operations within all possible programs, and therefore does not suggest a practical way to prove refinement between datatypes. The purpose of a rule for proving refinement is to allow each pair of corresponding operations to be considered in isolation, and not in the context of any program. This is achieved with the aid of a representation operation, which computes concrete states from abstract states. We say that the representation operation is a *simulation* between the two datatypes.

Definition 3 *An operation* (rep) *is a simulation from* $(I, \{O_j | j \in J\}, F)$ *to* $(I', \{O'_j | j \in J\}, F')$ *if the following inequations hold.*

$$
\begin{array}{lrcl}
(S1) & I \mathbin{\S} rep & \sqsubseteq & I' \\
(S2) & O_j \mathbin{\S} rep & \sqsubseteq & rep \mathbin{\S} O'_j \quad \text{for all } j \in J \\
(S3) & F & \sqsubseteq & rep \mathbin{\S} F'
\end{array}
$$

♡

Note that *rep* is another example of an operation for which execution begins and ends in different state spaces: it starts execution in the presence of global variables and concrete, local variables; it finishes execution in the presence of global variables and abstract, local variables. Again, this structure need not be explicit in the definition.

By reversing the direction of computation in Def. 3, a related set of inequations is obtained. We say these define a *cosimulation*.

Definition 4 *An operation* (rep) *is a cosimulation from* $(I, \{O_j | j \in J\}, F)$ *to* $(I', \{O'_j | j \in J\}, F')$ *if the following inequations hold.*

$$
\begin{array}{lrcl}
(CS1) & I & \sqsubseteq & I' \mathbin{\S} rep \\
(CS2) & rep \mathbin{\S} O_j & \sqsubseteq & O'_j \mathbin{\S} rep \quad \text{for } j \in J \\
(CS3) & rep \mathbin{\S} F & \sqsubseteq & F'
\end{array}
$$

♡

It has long been known that the construction of simulations, as in Def. 3, is a sound method for establishing refinement between datatypes [37], but that alone the method is not complete. Hoare et al [26] achieve completeness by formulating the above cosimulation rule and by using it together with simulation. Their work was based on a relational model of computation. In Sect. 4 and 5 we show that the use of predicate transformers makes the construction of cosimulations, alone, a sound and complete method.

3 Predicate transformers

The main result presented in this paper is that cosimulation alone gives a complete method for data refinement, provided one works with monotonic predicate transformers rather than relations. In this section we introduce some notation, and list the well-known properties on which the completeness result relies.

From now on, we consider a specific interpretation of programs, using weakest-precondition semantics: for program P and predicate ψ we denote by $P\,\psi$ the weakest precondition of P with respect to ψ, so effectively we identify each program with its meaning as a predicate transformer. We will need to consider several subclasses of predicate transformers defined by the following properties.

$$
\begin{array}{lll}
(strict) & P\,\bot = \bot & \\
(total) & P\,\top = \top & \\
(disjunctive) & P\,(\bigvee_{i\in I} \psi_i) = (\bigvee_{i\in I} P\,\psi_i) & \text{provided } I \neq \varnothing \\
(conjunctive) & P\,(\bigwedge_{i\in I} \psi_i) = (\bigwedge_{i\in I} P\,\psi_i) & \text{provided } I \neq \varnothing
\end{array}
$$

where $\top$ and $\bot$ are the predicates **true** and **false** respectively.

We interpret $\sqsubseteq$ as follows:

$$P \sqsubseteq Q \ \widehat{=}\ (\forall \psi \bullet P\,\psi \Rightarrow Q\,\psi).$$

As is well known, when programs are modelled using weakest-precondition semantics, this ordering naturally models refinement between programs [4, 44].

We write $\sqcap$ and $\sqcup$ for the meet and join with respect to $\sqsubseteq$.

$$
\begin{array}{lll}
(P \sqcap Q)\,\psi & \widehat{=} & (P\,\psi) \wedge (Q\,\psi) \\
(P \sqcup Q)\,\psi & \widehat{=} & (P\,\psi) \vee (Q\,\psi)
\end{array}
$$

$P \sqcap Q$ models the nondeterministic choice between P and Q. The operator $\sqcup$ has been used to make formal the use of logical variables [46], and to represent angelic choice [11].

We interpret $\mathbin{\fatsemi}$ as the backward composition of predicate transformers.

$$(P \mathbin{\fatsemi} Q)\,\psi \ \widehat{=}\ P\,(Q\,\psi)$$

We do this so that $\mathbin{\fatsemi}$ corresponds to sequential execution of programs when weakest-precondition semantics are used.

We write Id for the identity function on predicates. Id represents the null program, which leaves the values of variables unchanged.

We will first prove completeness and soundness in a rather restricted setting, with programs being total, conjunctive predicate transformers. The restriction on programs requires a correspondingly restricted programming language, so for now we consider a language $Tprog$, which constructs programs using only Id, $\fatsemi$, and $\sqcap$; each of these operators preserves totality and conjunctivity. $Tprog$ is not as restrictive as it may at first seem: since nonstrict programs are allowed, guards (denoted $[G]$) can be included among the operations of a datatype. Guards (called coercions in [46]) are defined as follows.

$$[G]\,\psi \;\widehat{=}\; G \Rightarrow \psi$$

Using guards, alternation can be represented; for example

if G **then** S **else** T **fi** $\widehat{=}\; [G]\fatsemi S \sqcap [\neg G]\fatsemi T.$

The restriction to conjunctive total programs allows us to use the well-known Galois connection between weakest precondition and strongest postcondition: for each conjunctive total predicate transformer P, there exists a disjunctive, strict predicate transformer $\overline{P}$ such that

$(G1)\quad Id \;\sqsubseteq\; P\fatsemi \overline{P}$
$(G2)\quad \overline{P}\fatsemi P \;\sqsubseteq\; Id.$

The standard construction of $\overline{P}$ is

$$\overline{P}\,\phi \;\widehat{=}\; \left(\bigwedge_{\phi \Rightarrow P\,\theta} \theta\right).$$

Back and von Wright [11] use this sort of weak inverse in program calculation. Our completeness result also relies heavily on its use; it is the absence of such a Galois connection that prevents a single complete rule using relations.

In Sect. 6 we drop the restriction to terminating programs, and include recursion in the language, thus making our result applicable to almost all imperative programming languages. We retain the restriction to nonangelic programs, but then very few programming languages provide angelic choice.

4 Completeness

To demonstrate completeness of cosimulation, we consider an arbitrary instance of refinement between abstract datatypes, (I, OP, F) is refined by (I', OP', F')

say, and construct a cosimulation between them. The required cosimulation, taking predicates over abstract state variables to predicates over concrete state variables, can be constructed explicitly in terms of the two datatypes. The intuition behind the construction is as follows. Consider a program that may use either the abstract datatype or the concrete, and imagine stopping this program during its execution: if it were using the abstract datatype then some sequence of operations ($I \mathbin{\raise.4ex\hbox{$\scriptscriptstyle\circ$}\kern-.2em\raise-.4ex\hbox{$\scriptscriptstyle\circ$}} S$ say, where S is sequence of operations drawn from OP) would have been applied; if it were using the concrete then a corresponding sequence of operations ($I' \mathbin{\raise.4ex\hbox{$\scriptscriptstyle\circ$}\kern-.2em\raise-.4ex\hbox{$\scriptscriptstyle\circ$}} S'$) would have been applied. One can calculate concrete predicates from abstract ones by first using weakest precondition to look backwards through $I \mathbin{\raise.4ex\hbox{$\scriptscriptstyle\circ$}\kern-.2em\raise-.4ex\hbox{$\scriptscriptstyle\circ$}} S$, and then using strongest postcondition to look forward through $I' \mathbin{\raise.4ex\hbox{$\scriptscriptstyle\circ$}\kern-.2em\raise-.4ex\hbox{$\scriptscriptstyle\circ$}} S'$. Lastly, to obtain a concrete predicate whose truth is independent of the stopping point, the disjunction over all sequences is taken. Hence we define

$$rep^\perp \psi = \left(\bigvee_{S \in OP^*} \overline{(I' \mathbin{\raise.4ex\hbox{$\scriptscriptstyle\circ$}\kern-.2em\raise-.4ex\hbox{$\scriptscriptstyle\circ$}} S')} \, (I \mathbin{\raise.4ex\hbox{$\scriptscriptstyle\circ$}\kern-.2em\raise-.4ex\hbox{$\scriptscriptstyle\circ$}} S) \, \psi \right),$$

where OP^* is the closure of OP under sequential composition.

Theorem 1 *For total, conjunctive operations and for language Tprog, if (I, OP, F) is refined by (I', OP', F') then $rep^\perp$ is a cosimulation between them.*

Proof

Note that the elements of OP^* are programs constructed from the elements of OP, and so by Def. 2

(R1) $(I \mathbin{\raise.4ex\hbox{$\scriptscriptstyle\circ$}\kern-.2em\raise-.4ex\hbox{$\scriptscriptstyle\circ$}} S \mathbin{\raise.4ex\hbox{$\scriptscriptstyle\circ$}\kern-.2em\raise-.4ex\hbox{$\scriptscriptstyle\circ$}} F) \sqsubseteq (I' \mathbin{\raise.4ex\hbox{$\scriptscriptstyle\circ$}\kern-.2em\raise-.4ex\hbox{$\scriptscriptstyle\circ$}} S' \mathbin{\raise.4ex\hbox{$\scriptscriptstyle\circ$}\kern-.2em\raise-.4ex\hbox{$\scriptscriptstyle\circ$}} F')$ for each $S \in OP^*$

From this we derive the three cosimulation properties

(CS1)

$$\begin{aligned}
&\quad I \, \psi \\
\Rightarrow &\quad I' \, \overline{I'} \, I \, \psi &&\text{G1}\\
\Rightarrow &\quad I' \left(\bigvee_{S \in OP^*} \overline{(I' \mathbin{\raise.4ex\hbox{$\scriptscriptstyle\circ$}\kern-.2em\raise-.4ex\hbox{$\scriptscriptstyle\circ$}} S')} \, (I \mathbin{\raise.4ex\hbox{$\scriptscriptstyle\circ$}\kern-.2em\raise-.4ex\hbox{$\scriptscriptstyle\circ$}} S) \, \psi \right) &&\text{Id} \in OP^*, \text{monotonicity of } I' \\
= &\quad I' \, rep^\perp \, \psi &&\text{definition}
\end{aligned}$$

(CS2)

$$\begin{aligned}
&\quad rep^\perp \, O \, \psi \\
= &\quad \left(\bigvee_{S \in OP^*} \overline{(I' \mathbin{\raise.4ex\hbox{$\scriptscriptstyle\circ$}\kern-.2em\raise-.4ex\hbox{$\scriptscriptstyle\circ$}} S')} \, (I \mathbin{\raise.4ex\hbox{$\scriptscriptstyle\circ$}\kern-.2em\raise-.4ex\hbox{$\scriptscriptstyle\circ$}} S) \, O \, \psi \right) &&\text{definition}
\end{aligned}$$

$$
\begin{aligned}
&\Rightarrow && \left(\bigvee\nolimits_{S \in OP^\bullet} O' \; \overline{O' \; \overline{(I' \; \mathring{,} \; S')}} \; (I \; \mathring{,} \; S) \; O \; \psi\right) && G1 \\
&= && \left(\bigvee\nolimits_{S \in OP^\bullet} O' \; \overline{(I' \; \mathring{,} \; S' \; \mathring{,} \; O')} \; (I \; \mathring{,} \; S \; \mathring{,} \; O) \; \psi\right) && \text{since } \overline{P \; \mathring{,} \; Q} = \overline{Q} \; \mathring{,} \; \overline{P} \\
&\Rightarrow && \left(\bigvee\nolimits_{T \in OP^\bullet} O' \; \overline{(I' \; \mathring{,} \; T')} \; (I \; \mathring{,} \; T) \; \psi\right) && \text{closure} \\
&\Rightarrow && O' \; \left(\bigvee\nolimits_{T \in OP^\bullet} \overline{(I' \; \mathring{,} \; T')} \; (I \; \mathring{,} \; T) \; \psi\right) && \text{monotonicity of } O' \\
&= && O' \; rep^\perp \; \psi && \text{definition}
\end{aligned}
$$

(CS3)

$$
\begin{aligned}
&&& rep^\perp \; F \; \psi \\
&= && \left(\bigvee\nolimits_{S \in OP^\bullet} \overline{(I' \; \mathring{,} \; S')} \; (I \; \mathring{,} \; S) \; F \; \psi\right) && \text{definition} \\
&= && \left(\bigvee\nolimits_{S \in OP^\bullet} \overline{(I' \; \mathring{,} \; S')} \; (I \; \mathring{,} \; S \; \mathring{,} \; F) \; \psi\right) && \text{definition of } \mathring{,} \\
&\Rightarrow && \left(\bigvee\nolimits_{S \in OP^\bullet} \overline{(I' \; \mathring{,} \; S')} \; (I' \; \mathring{,} \; S' \; \mathring{,} \; F') \; \psi\right) && R1 \\
&= && \left(\bigvee\nolimits_{S \in OP^\bullet} \overline{(I' \; \mathring{,} \; S')} \; (I' \; \mathring{,} \; S') \; F' \; \psi\right) && \text{definition of } \mathring{,} \\
&\Rightarrow && \left(\bigvee\nolimits_{S \in OP^\bullet} F' \; \psi\right) && G2 \\
&= && F' \; \psi && \text{property of } \bigvee
\end{aligned}
$$

♡

The proof just given can be presented far more cleanly, although very abstractly, in terms of Category Theory [18]. As such the proof shows that certain lax natural transformations can be extended.

In our proof of completeness we have exhibited only one cosimulation. In fact there may be many. The set of cosimulations between any two datatypes is not without structure. The set forms a lattice with respect to $\sqsubseteq$: if $rep1$ and $rep2$ are cosimulations, then so are $rep1 \sqcup rep2$ and $rep1 \sqcap rep2$. In fact, the set is closed under nonempty meets and joins of arbitrary size. The bottom of the lattice is $rep^\perp$, and the top can be defined similarly as follows.

$$rep^\top \; \psi = \left(\bigwedge\nolimits_{S \in OP^\bullet} (S' \; \mathring{,} \; F') \; \overline{(S \; \mathring{,} \; F)} \; \psi\right)$$

5 Soundness

To demonstrate soundness of the cosimulation rule, we must consider an arbitrary pair of datatypes (I, OP, F) and (I', OP', F') for which a cosimulation rep exists, and then show that (I', OP', F') does indeed refine (I, OP, F).

Theorem 2 *For language Tprog, if there is a cosimulation (rep) between (I, OP, F) and (I', OP', F'), then (I, OP, F) is refined by (I', OP', F').*

Proof

First we show that, for every function $\mathcal{P}$, expressible as a composition of program-language operators,

$$(CSn) \quad rep \mathbin{\raisebox{0.5ex}{,}\mkern-3mu\raisebox{-0.5ex}{,}} \mathcal{P}(OP) \sqsubseteq \mathcal{P}(OP') \mathbin{\raisebox{0.5ex}{,}\mkern-3mu\raisebox{-0.5ex}{,}} rep$$

Base case:
We need consider only programs that consist of a call to a single operation. In this case (CSn) follows directly from (CS2).

Case (Id):

$$\begin{array}{lll}
& rep \mathbin{;} Id & \\
= & rep & identity \\
= & Id \mathbin{;} rep & identity
\end{array}$$

Case ($\mathbin{;}$):

$$\begin{array}{lll}
& rep \mathbin{;} (P \mathbin{;} Q) & \\
\sqsubseteq & P' \mathbin{;} rep \mathbin{;} Q & Inductive\ assumption \\
\sqsubseteq & (P' \mathbin{;} Q') \mathbin{;} rep & Inductive\ assumption
\end{array}$$

Case ($\sqcap$):

$$\begin{array}{lll}
& rep \mathbin{;} (P \sqcap Q) & \\
\sqsubseteq & (rep \mathbin{;} P) \sqcap (rep \mathbin{;} Q) & Monotonicity\ of\ rep \\
\sqsubseteq & (P' \mathbin{;} rep) \sqcap (Q' \mathbin{;} rep) & Inductive\ assumption \\
= & (P' \sqcap Q') \mathbin{;} rep & Property\ of\ \sqcap
\end{array}$$

This concludes the proof of (CSn). The theorem now follows as below.

$$\begin{array}{lll}
& I \mathbin{;} \mathcal{P}(OP) \mathbin{;} F & \\
\sqsubseteq & I' \mathbin{;} rep \mathbin{;} \mathcal{P}(OP) \mathbin{;} F & CS1 \\
\sqsubseteq & I' \mathbin{;} \mathcal{P}(OP') \mathbin{;} rep \mathbin{;} F & CSn \\
\sqsubseteq & I' \mathbin{;} \mathcal{P}(OP') \mathbin{;} F' & CS3
\end{array}$$

♡

6 Partial programs

In this section we generalise the completeness and soundness results to programs that may fail to terminate. Thus programs are modeled by conjunctive, but not-necessarily-total predicate transformers. We also enrich the language *Tprog*, adding recursion and the aborting program. We call the enriched language *Prog*.

Recursion is modeled by least fixed point with respect to $\sqsubseteq$, written ($\mu X \bullet \mathcal{P}(X)$), where $\mathcal{P}$ is a function, expressible as a composition of program-language operators. The aborting program, written 0 is defined by

$$0 \, \psi = \bot.$$

The language *Prog* consists of all the programs constructions using Id, $\mathring{,}$, $\sqcap$, 0, and μ; each of these constructs preserves conjunctivity.

Now that nonterminating programs are accepted, we can no longer assume the Galois connection between weakest precondition and strongest postcondition, but we can borrow the generalised inverse of Bach and von Wright [11]. For each conjunctive predicate transformer P, Bach and von Wright define a strict, disjunctive predicate transformer $\overline{P}$, satisfying the following weaker inequations.

$$\begin{array}{ll}(G1') & \{P \, \top\} \sqsubseteq P \mathring{,} \overline{P} \\ (G2') & \overline{P} \mathring{,} P \sqsubseteq Id,\end{array}$$

where $\{\phi\}$ is an *assertion*, defined by

$$\{\phi\} \, \psi \; \widehat{=} \; \phi \wedge \psi.$$

Assertions have the following easily derived properties.

(A1) if $\phi \Rightarrow \psi$ then $\{\phi\} \sqsubseteq \{\psi\}$

(A2) $P = \{P \, \top\} \mathring{,} P$

From these properties we can derive conditions under which a use of $(G1')$ can replace a use of $(G1)$.

Lemma 1 *For predicate transformer P and conjunctive predicate transformer Q, if $P \, \top \Rightarrow Q \, \top$, then*

$$P \sqsubseteq Q \mathring{,} \overline{Q} \mathring{,} P$$

Proof

$$\begin{aligned} & P \\ =\ & \{P\ \top\}\ \mathring{,}\ P & & A2 \\ \sqsubseteq\ & \{Q\ \top\}\ \mathring{,}\ P & & A1 \\ \sqsubseteq\ & Q\ \mathring{,}\ \overline{Q}\ \mathring{,}\ P & & G1' \end{aligned}$$

♡

In this more general setting, where nontermination is permitted, we must impose some extra conditions on the datatypes and on the cosimulation.

- The datatypes are required to have strict, total finalisations. This means that the finalisations must be guaranteed to terminate and must not be guarded. Both these assumptions are quite reasonable, since finalisations are typically just projections from the product of the global and local state spaces into the global space, as we explained in Sect. 2.

- The cosimulation must be strict and continuous (that is, disjunctive for chains). This condition is required for cosimulation to be sound in the presence of recursion.

- The abstract operations are required to be continuous. In the proof of completeness, this is required when establishing the continuity of the cosimulation. This last constraint is very undesirable; it says that the abstract operations are required to be only boundedly nondeterministic, and therefore rules out some simple mathematical specifications, such as "Pick any integer". We know of no sound and complete data refinement methods that allow both recursion and unbounded nondeterminism.

Theorem 3 *For conjunctive, continuous operations and for language Prog, if (I, OP, F) is refined by (I', OP', F') then $rep^\perp$ is a strict, continuous cosimulation.*

Proof

To prove that $rep^\perp$ is a cosimulation, we analyse the proof of Theorem 1 and check that each use of $(G1)$ satisfies the condition of Lemma 1. There are two such uses. The first is in establishing $I \sqsubseteq I'\ \mathring{,}\ \overline{I'}\ \mathring{,}\ I$. Use of Lemma 1 requires $I\ \top \Rightarrow I'\ \top$, which can be derived as follows.

$$\begin{aligned} & I\ \top \\ =\ & I\ F\ \top & & F\text{ is total} \\ \Rightarrow\ & I'\ F'\ \top & & R1,\text{ see Theorem 1} \\ =\ & I'\ \top & & F'\text{ is total} \end{aligned}$$

The second use of (G1) is in establishing

$$\overline{(I' \mathbin{\raisebox{0.5ex}{,}\!\raisebox{-0.5ex}{,}} S')} \mathbin{\raisebox{0.5ex}{,}\!\raisebox{-0.5ex}{,}} (I \mathbin{\raisebox{0.5ex}{,}\!\raisebox{-0.5ex}{,}} S) \mathbin{\raisebox{0.5ex}{,}\!\raisebox{-0.5ex}{,}} O \sqsubseteq O' \mathbin{\raisebox{0.5ex}{,}\!\raisebox{-0.5ex}{,}} \overline{O'} \mathbin{\raisebox{0.5ex}{,}\!\raisebox{-0.5ex}{,}} \overline{(I' \mathbin{\raisebox{0.5ex}{,}\!\raisebox{-0.5ex}{,}} S')} \mathbin{\raisebox{0.5ex}{,}\!\raisebox{-0.5ex}{,}} (I \mathbin{\raisebox{0.5ex}{,}\!\raisebox{-0.5ex}{,}} S) \mathbin{\raisebox{0.5ex}{,}\!\raisebox{-0.5ex}{,}} O.$$

Here, use of Lemma 1 requires

$$\overline{(I' \mathbin{;} S')} \, (I \mathbin{;} S) \, O \, \top \Rightarrow O' \, \top,$$

which can be derived similarly.

$$
\begin{array}{rll}
 & \overline{(I' \mathbin{;} S')} \, (I \mathbin{;} S) \, O \, \top & \\
= & \overline{(I' \mathbin{;} S')} \, (I \mathbin{;} S) \, O \, F \, \top & F \text{ is total} \\
\Rightarrow & \overline{(I' \mathbin{;} S')} \, (I' \mathbin{;} S') \, O' \, F' \, \top & R1 \\
\Rightarrow & O' \, F' \, \top & G2 \\
= & O' \, \top & F' \text{ is total}
\end{array}
$$

We must also show that $rep^\perp$ is strict and continuous. Strictness follows from $\overline{(I' \mathbin{;} S')} \mathbin{;} (I \mathbin{;} S)$ being strict for each $S \in OP*$, which can be demonstrated as follows.

$$
\begin{array}{rll}
 & \overline{(I' \mathbin{;} S')} \, (I \mathbin{;} S) \, \perp & \\
= & \overline{(I' \mathbin{;} S')} \, (I \mathbin{;} S) \, F \, \perp & F \text{ is strict} \\
\Rightarrow & \overline{(I' \mathbin{;} S')} \, (I' \mathbin{;} S') \, F' \, \perp & R1 \\
\Rightarrow & F' \, \perp & G2 \\
= & \perp & F' \text{ is strict}
\end{array}
$$

Continuity follows from our assumption that the abstract operations are continuous, and from $\overline{(I' \mathbin{;} S')}$ being disjunctive (and therefore continuous).

♡

Theorem 4 *For language Prog, if there is a strict, continuous cosimulation rep between (I, OP, F) and (I', OP', F'), then (I, OP, F) is refined by (I', OP', F').*

Proof

The proof of Theorem 2 still applies here, provided that the two extra constructs (O and μ) are considered. The Limit Theorem (as generalized by Hitchcock and Park) asserts that, for monotonic $\mathcal{P}$, there exists an ordinal γ such that $(\mu X \bullet \mathcal{P}(X)) = \mathcal{P}^\gamma$, where

$$
\begin{array}{rcl}
\mathcal{P}^0 & = & 0 \\
\mathcal{P}^{\alpha+1} & = & \mathcal{P}(\mathcal{P}^\alpha) \\
\mathcal{P}^\lambda & = & (\bigsqcup_{\beta < \lambda} \mathcal{P}^\beta)
\end{array}
$$

Hence, recursion adds only 0 and joins of chains to the language.

Case (0):

$$\begin{aligned}
&\quad rep\ 0\ \phi \\
&= rep\ \bot &&\text{definition of } 0\\
&= \bot &&\text{rep is strict}\\
&= 0\ rep\ \bot &&\text{definition of } 0
\end{aligned}$$

Case ($\bigsqcup$ of chains):

$$\begin{aligned}
&\quad rep\ (\bigsqcup_{i \in I} P_i)\ \phi \\
&= rep\ (\bigvee_{i \in I} P_i\ \phi) &&\text{definition of } \bigsqcup \\
&= (\bigvee_{i \in I} rep\ P_i\ \phi) &&\text{rep is continuous} \\
&= (\bigvee_{i \in I} P'_i\ rep\ \phi) &&\text{inductive assumption} \\
&= (\bigsqcup_{i \in I} P'_i)\ rep\ \phi &&\text{definition of } \bigsqcup
\end{aligned}$$

♡

For partial programs, as for total programs, the set of cosimulations forms a lattice with respect to $\sqsubseteq$. The set is closed under arbitrary nonempty joins but only finite nonempty meets. The bottom of the lattice is $rep^\bot$. There may be no top.

7 An example

Recall from the introduction that both [32] and [1] show that the standard methods [30, 33] of data refinement are incomplete. Both authors exhibit incompleteness with example refinements that cannot be proved using the standard methods. Here we show how our single rule establishes one of their example refinements (see Figures 1 and 2). For comparison we have based the example on the state machines of [1, Sec. 5.2].

Module A of Figure 1 contains two variables, both of type $0..10$, so that i and j constitute the abstract state space. Variable i is *exported*, which means it can be accessed but not modified by programs using the module. The initialisation sets i to 0 but does not constrain j except to choose it from its type (so that $0 \leq j \leq 10$).

If j is initialised to J, say, then the first J calls of procedure *Inc* will increment i but subsequent calls will have no effect. The nondeterministic initialisation of j determines *in advance* how many increments can occur; but that information is not available externally because j is not exported. Following [1], variable j would be called a *prophesy* variable.

module A
  export i, Inc;
  var $i, j: 0..10$;

  procedure $Inc \;\hat{=}\;$ **if** $j > 0 \rightarrow i, j := i+1, j-1$
  $\quad\quad\quad\quad\quad\quad\quad\;$ [] $\;j = 0 \rightarrow$ **skip**
  $\quad\quad\quad\quad\quad\quad\;\;$ **fi**;

  initially $i = 0$
end

Figure 1: Module with 'prophesy' variable j

module C
  export i, Inc;
  var $i: 0..10;\;\; b: Boolean$;

  procedure $Inc \;\hat{=}\;$ **if** $\neg b \wedge i < 10 \rightarrow i := i+1$
  $\quad\quad\quad\quad\quad\quad\quad\;$ [] $\;true \quad\quad\quad \rightarrow b := true$
  $\quad\quad\quad\quad\quad\quad\;\;$ **fi**;

  initially $i = 0$
end

Figure 2: Module with j replaced by Boolean variable b

In Module C (concrete) of Figure 2, variable j has been replaced by a Boolean variable b; thus the concrete state space comprises i and b. While b is false, procedure *Inc* might or might not increment i; but once b becomes true, again subsequent calls have no effect. The initialisation does not constrain b.

Neither A nor C can increment i beyond 10, and in fact their external behaviours are identical: no program can distinguish one module from the other, since j and b are inaccessible.

But, as [1] remarks, the standard methods (for example [23, 51, 47]) fail to show that A is refined by C, because the nondeterministic choice in A is made earlier than in C — a phenomenon well-known to be associated with such failures. Those methods are equivalent in power [16], and the time-honoured *auxiliary variable technique* is equivalent also [41] (and so fails as well).

But we can apply our rule to refine A to C if we take *rep* as follows:

$$rep\ \phi \ \widehat{=}\ (\forall j :: (b \Rightarrow j = 0) \wedge i + j \leq 10 \Rightarrow \phi).$$

This predicate transformer, transforming a predicate over the abstract space into one over the concrete state space, is not of the form allowed by the standard methods, where *rep* must distribute disjunction [20, 47].

Def. 4 then requires that we show, for all abstract ϕ (over i, j but not b), that

$$\begin{aligned}&(\forall j :: (b \Rightarrow j = 0) \wedge i + j \leq 10 \Rightarrow wp(Inc_A, \phi)) \\ \Rightarrow\ & wp(Inc_C, (\forall j :: (b \Rightarrow j = 0) \wedge i + j \leq 10 \Rightarrow \phi)).\end{aligned}$$

Proof of the above is straightforward if lengthy.

There is also a proof obligation for the initialisation, but it is trivial in this case: we must show that

$$i = 0 \ \Rightarrow\ (\forall j :: (b \Rightarrow j = 0) \wedge i + j \leq 10 \Rightarrow i = 0).$$

Section 6 showed that if we wish to exchange module A for C in the presence of recursion, then we require additionally that *rep* be strict and continuous; it is in fact since for all $i: 0..10$ and b: *Boolean* there are a nonzero, but-only-finite number of values for j satisfying the predicate $(b \Rightarrow j = 0) \wedge i + j \leq 10$. (Remember that the type of i guarantees $0 \leq i \leq 10$.)

8 Conclusion

Although the completeness of a single method is of considerable interest theoretically, there may still be practical advantages in retaining two forms. For example, both [51, 47] show that (standard) simulation can be applied directly to the predicates in the text of an abstract program, producing a concrete program by calculation. For cosimulation, the prophesy technique is simple and easy to understand.

It is interesting to note that a transformer is both a simulation and a cosimulation exactly when it is functional — when it is a refinement *mapping*, in the sense of [30, 33, 1].

Abadi and Lamport's method [1] treats infinite behaviours and internal transitions (stuttering), and thus its application is more general than we have considered here. In our simpler context, however, we have shown that explicit use of history and prophesy is unnecessary. It is known already that history variables are subsumed by the use of refinement relations rather than functions [33, p.237]. We have shown that both history and prophesy variables are subsumed by the use of predicate transformers rather than relations.

Back [8] uses generalised program fragments for data refinement. Thus from a semantic point of view, he too uses predicate transformers rather than relations. He identifies [8, p.15] two approaches to data refinement, and chooses the 'more general'; his Theorem 1 then establishes equivalence of that to downwards simulation, with a small (and unnecessary) restriction to avoid miracles. It is possible that Back's other approach is upward simulation, and that completeness is within his reach.

We do not know of any method that reasons over program fragments, rather than entire behaviours, for which bounded-nondeterminism does not limit completeness.

9 Acknowledgement

We are grateful for Jim Woodcock's help in finding the predicate-transforming version of upward simulation, and we thank Tony Hoare for his comments.

Types and Invariants in the Refinement Calculus

Carroll Morgan and Trevor Vickers

Abstract

A rigorous treatment of types as sets is given for the refinement calculus, a method of imperative program development. It is simple, supports existing practice, casts new light on type-checking, and suggests generalisations that might be of practical benefit. Its use is illustrated by example.

1 Introduction

Program developments in the style of Dijkstra [17] rely on *implicit* typing of variables. One agrees beforehand that all variables have a certain type (say Z, the integers); then individual steps are justified by referring to that type, where necessary. For example, the truth of the entailment

$$a < b \Rightarrow wp(a := a+1, a \leq b)$$

$\Leftarrow$ depends on a, b being integers. But that dependence is at present informal.

In the refinement calculus also [5, 48, 50], the dependence is informal; and the contribution of this paper is to make it rigorous. We make *typed* local variable declarations affect the meaning of commands within their scope, and allow development steps there to refer to that type information.

In fact, typing is a special kind of invariant: in the scope of the declaration **var** $n: N$, which introduces a new local variable n of type N (the natural numbers), the invariant is $n \in N$ and all commands preserve it. We allow the declaration of *local invariants* in general, and the rules for typing follow from that.

A surprising feature of our approach is that imposing an invariant does not increase the developer's proof obligations in the usual (prohibitive) way: it is

---

An expansion of [45].
Appeared in Science of Computer Programming 14 (1990). (North Holland)

$$wp_I(\mathbf{skip}, \phi) \;\widehat{=}\; I \wedge \phi$$
$$wp_I(\mathbf{abort}, \phi) \;\widehat{=}\; \mathit{false}$$
$$wp_I(x := E, \phi) \;\widehat{=}\; I \wedge (I \Rightarrow \phi)[x \backslash E]$$
$$wp_I(P; Q, \phi) \;\widehat{=}\; wp_I(P, wp_I(Q, \phi))$$

$$wp_I(\mathbf{if}\ ([]i \bullet G_i \to P_i)\ \mathbf{fi}, \phi)$$
$$\widehat{=}\ (\vee i \bullet G_i)$$
$$\wedge\ (\wedge i \bullet G_i \Rightarrow wp_I(P_i, \phi))$$

The substitution $[x \backslash E]$ replaces all free occurrences of x by E, with suitable renaming of bound variables if necessary to avoid capture. Iteration **do** $\cdots$ **od**, a special case of recursion, is dealt with in Section 8.3.

Figure 1: Invariant semantics for Dijkstra's language

not necessary to prove, during development, that the invariant is maintained. It is maintained automatically.

Thus even programs that appear to break the invariant actually maintain it: instead of being type-incorrect, they are miracles [44, 50, 54]. But miracles are still programs, and that allows a more uniform calculus of program refinement.

Nevertheless, a check is necessary to exclude miracles from the final program, since they cannot be executed. That check, like the type checking which it subsumes, is often obvious and can in many cases be delegated to machine. When a machine cannot perform it, it is only because the program developer has used more general invariants than typing.

We believe that it is important to separate the use of an invariant (or type) from the proof that it is respected. Continual formal type-checking *during* development is impractical — and that has, so far, limited the rigorous use of types in imperative programs.

This paper extends the earlier [45] with the examples of Section 9.

2 Invariant semantics

We retain Dijkstra's language, but now give its meaning relative to an invariant, which we call the *context*. Any formula over the program variables (even *false*) may be a context. We write $wp_I(P, \phi)$ for the weakest precondition in context I of a program P with respect to a postcondition ϕ, and give the resulting semantics of Dijkstra's language in Figure 1. Note that taking I to be *true* in Figure 1 gives the usual semantics: therefore we say that *true* is the default context.

These lemmas support our choice of semantics in Figure 1:

Lemma 1 *Assume invariant.* No program P, in context I, is guaranteed to terminate unless I holds initially:

$$wp_I(P, true) \Rrightarrow I.$$

Proof: Structural induction over P.
♡

Lemma 2 *Establish invariant.* Any program P, in context I, establishes I iff it establishes anything:

$$wp_I(P, \phi) \equiv wp_I(P, I \wedge \phi).$$

Proof: Structural induction over P.
♡

Note that $wp_I(P, \sqcup)$ is still monotonic over implication (in its second argument), and still distributes conjunction. Note also that, in the (typing) context $n \in \mathbf{N}$, the command $n := -1$ terminates, and hence (Lemma 2) re-establishes the invariant! We return to that later.

3 The refinement calculus

The refinement calculus is based on an extended programming language, in which specifications can be written, and a relation of refinement between its programs such that implementations refine their specifications [5, 48, 50].

3.1 Language extensions

A *specification* is a list w of changing variables, called the *frame*, and a formula *post*, called the *postcondition*. It is defined as follows:

Definition 1 *Specification.*

$$wp_I(w\colon [post], \phi) \quad \widehat{=} \quad I \wedge (\forall w \bullet I \wedge post \Rightarrow \phi).$$

♡

Quantifications are written within parentheses $(\cdots)$, and the bound variable list is terminated by •. Our precedence for propositional connectives is (highest) $\neg, \wedge, \vee, \Rightarrow, \Leftrightarrow$ (lowest).

In the special case of a specification with an empty list of changing variables, we have a *coercion*, whose definition is derived from Definition 1:

Definition 2 *Coercion.*

$$wp_I([post], \phi) \;\;\widehat{=}\;\; I \wedge (post \Rightarrow \phi).$$

♡

An *assertion* is a single condition *pre*, written $\{pre\}$, and is defined as follows:

Definition 3 *Assertion.*

$$wp_I(\{pre\}, \phi) \;\;\widehat{=}\;\; I \wedge pre \wedge \phi.$$

♡

Coercions and assertions are together known as *annotations*. We omit the sequential composition operator ; whenever one or both of its arguments is an annotation.

Assertions and specifications often occur together; for example, the program

$$\{m > 0\}\; n\colon [n < m] \tag{1}$$

sets n below m provided m was positive to begin with. (Note that Definitions 1 and 3 differ slightly from the notation of [44], where Specification (1) would be written $n\colon [m > 0, n < m]$. The present notation agrees with [50]; but [5] remains substantially different.)

An untyped local variable x is introduced using the declaration **var** and scope brackets $|[\cdots]|$. It is defined

Definition 4 *Untyped local variable.* Provided neither I nor ϕ contains free x,

$$wp_I(|[\,\mathbf{var}\; x \bullet P\,]|, \phi) \;\;\widehat{=}\;\; (\forall x \bullet wp_I(P, \phi)).$$

♡

Definition 4 is standard; later, we extend it for typed local variables.

A local invariant J is introduced by the declaration **inv** and scope brackets. It is defined

Definition 5 *Local invariant.*

$$wp_I(|[\textbf{ inv } J \bullet P \,]|, \phi) \quad \widehat{=} \quad wp_{I \wedge J}(P, \phi).$$

♡

In $|[\textbf{ inv } J \bullet P \,]|$, the new invariant J is assumed initially, is maintained automatically by every command in P, and therefore is established finally.

Finally, typed local variables are a combination of the above: an untyped declaration, an initialisation, and a local invariant. We have

Definition 6 *Typed local variables.* For any set T and formula I,

$$|[\textbf{ var } x \colon T \textbf{ and } I \bullet P \,]|$$
$$\widehat{=} \quad |[\textbf{ var } x \bullet x \colon [x \in T \wedge I]; \\ |[\textbf{ inv } x \in T \wedge I \bullet P \,]| \\ ]|.$$

♡

The "$:T$" gives the type of x; the "**and** I" gives an invariant to be imposed additionally. An invariant *true* may be omitted.

Like **inv**, the declaration **and** may appear on its own, so that

$$|[\textbf{ var } x \colon T \textbf{ and } I \bullet \cdots]|$$

is equivalent to the nested declarations

$$|[\textbf{ var } x \colon T \bullet |[\textbf{ and } I \bullet \cdots]| \,]|.$$

The difference between **inv** and **and** is only that the latter includes an initialisation.

Invariants introduced by **inv** or **and** are *explicit*; invariants introduced by typing $x \colon T$ are *implicit*.

3.2 The refinement relation

The refinement relation $\sqsubseteq$ holds between programs P and Q whenever Q satisfies every specification that P does, in every context. This is its definition:

Definition 7 *Refinement.* For programs P and Q, we have $P \sqsubseteq Q$ iff for all contexts I and postconditions ϕ,

$$wp_I(P, \phi) \Rightarrow wp_I(Q, \phi).$$

♡

Examples of refinement are given in the laws of Section 4.

If a program fragment is contained within the scope of a local invariant I, we can take advantage of the invariant by exploiting a weaker refinement relation $\sqsubseteq_I$, defined as follows:

Definition 8 *Refinement in context.* For programs P and Q,

$$P \sqsubseteq_I Q \quad \widehat{=} \quad |[\textbf{ inv } I \bullet P]| \sqsubseteq |[\textbf{ inv } I \bullet Q]|.$$

♡

In Definition 8, the index I of $\sqsubseteq_I$ is the context in which the refinement is valid. For example, with $I \widehat{=} n \in \mathbf{N}$, we have

$$x \colon [x \geq 0] \quad \sqsubseteq_I \quad x := n.$$

That is, setting x to a natural number n refines setting it to any non-negative value. (Law 4 below supplies an easy proof.)

In practice, we use the following lemma to demonstrate refinements in context:

Lemma 3 *Refinement in context.* For programs P and Q, we have $P \sqsubseteq_I Q$ iff for all postconditions ϕ, and stronger contexts J with $J \Rightarrow I$,

$$wp_J(P, \phi) \quad \Rightarrow \quad wp_J(Q, \phi).$$

Proof: From Definitions 8, 7, and 5, $P \sqsubseteq_I Q$ iff for all postconditions ϕ and contexts K

$$wp_{I \wedge K}(P, \phi) \quad \Rightarrow \quad wp_{I \wedge K}(Q, \phi).$$

But the set of contexts "$I \wedge K$ for all K" is exactly the set of contexts "J with $J \Rightarrow I$", as required.
♡

An immediate consequence of Lemma 3 is that strengthening the context cannot invalidate a refinement:

Lemma 4 *Strengthen context.* If $P \sqsubseteq_I Q$ and $J \Rightarrow I$, then also $P \sqsubseteq_J Q$.

Proof: Trivial, since Lemma 3 treats all stronger J than I.
♡

4 A development method

We regard as our "programming" language all the constructions of Figure 1 (traditional) and Section 3.1 (novel). That includes abstract programs, like

$$x \colon \left[x^2 - 3x + 2 = 0 \right],$$

and it of course admits "ordinary" programs too, like $x := 1$ and $x := 2$ (both of which refine the above).

Though "programming" and "ordinary" we use informally, we do precisely identify a subset of the language, called *code*, that can be executed automatically by computer. It includes all of Figure 1, and Definitions 3 and 6 without **and** of Section 3.1. (But in the case of Definition 6, we do restrict the language — in code — with which the set T can be expressed.) Code does *not* include specifications, coercions, untyped local variables, or explicit invariants.

In the refinement calculus, the aim of development is to refine a given program into another one written entirely in code. For that we use refinement laws, several of which are shown in this section. We need not use *wp* directly — it is used only to prove the refinement laws themselves.

For example, the following three laws deal with the use of context for simplification, and are easy consequences of Definition 7:

Law 1 *Weaken assumption.* Provided $I \wedge pre \Rightarrow pre'$,

$\{pre\} \quad \sqsubseteq_I \quad \{pre'\}$.

♡

Law 2 *Strengthen postcondition.* Provided $I \wedge post' \Rightarrow post$,

$$w\colon [post] \quad \sqsubseteq_I \quad w\colon [post'].$$

♡

Law 3 *Annotations and* **skip**.

$$\{pre\} \quad \sqsubseteq_I \quad \textbf{skip} \quad \sqsubseteq_I \quad [post].$$

♡

Law 3 means, for example, that at any point an assumption may be removed or a coercion inserted. But the above laws do not directly make progress towards code — which is after all our overall goal.

There are two defining aspects of code: first, any program written in code must be executable; and second, it must be decidable whether or not any text *is* code. The first aspect is straightforward: the language of code was *designed* for execution. Specifications, however, were not — and, in general, they *are* not, because most of the types of interest do not have recursively axiomatizable theories. Assertions, however, are code because they are refined by **skip** (Law 3).

The second aspect of code is straightforward, too, but requires "type checking". Consider this program:

$$|[\ \textbf{var}\ n\colon \mathbf{N} \bullet n := -1\ ]|. \tag{2}$$

Definitions 6, 4, 1, and 5 show that it does terminate in the default context *true*. But it establishes *false*:

$$\begin{aligned}
&\ wp(|[\ \textbf{var}\ n\colon \mathbf{N} \bullet n := -1\ ]|, false) \\
\equiv &\ (\forall n \bullet wp(n\colon [n \in \mathbf{N}], wp_{n \in \mathbf{N}}(n := -1, false))) \\
\equiv &\ (\forall n \bullet (\forall n \bullet n \in \mathbf{N} \Rightarrow n \in \mathbf{N} \wedge (-1 \in \mathbf{N} \Rightarrow false))) \\
\equiv &\ true.
\end{aligned}$$

Thus Program (2) violates Dijkstra's *Law of the excluded miracle* [17, p.18], and it cannot, therefore, be code.

That program is not code because it is ill-typed; but we defer the recognition of code to Section 7. Assuming we *can* recognise code, the development method is this: given a program, find code that refines it. Instrumental in that process are laws of refinement that, like the following, introduce code:

Law 4 *Assignment.* Provided $I \land pre \Rrightarrow post[w\backslash E]$,

$$\{pre\}\, w, x\!:\, [post] \quad \sqsubseteq_I \quad \{pre\}\, w\!:=\, E.$$

Proof: We use Lemma 3: suppose $J \Rrightarrow I$. Then

$$\begin{aligned}
&\quad wp_J(\{pre\}\, w, x\!:\, [post], \phi) \\
\equiv\ &\quad J \land pre \land J \land (\forall\, w, x\ \bullet\ J \land post \Rightarrow \phi) \\
\Rightarrow\ &\quad J \land pre \land (J[w\backslash E] \land post[w\backslash E] \Rightarrow \phi[w\backslash E]) \\
\Rightarrow\ &\quad \text{``assumptions''} \\
&\quad J \land pre \land (J \Rightarrow \phi)[w\backslash E] \\
\equiv\ &\quad wp_J(\{pre\}\, w\!:=\, E, \phi).
\end{aligned}$$

♡

The variables x in the frame, if any, are those that the assignment declines to change. The assumption $\{pre\}$ may be removed with Law 3, leaving only the assignment; but in Section 9 we see that sometimes it is best to leave the assumption there.

An important feature of Law 4 is that the context I plays a *constructive* role: the stronger it is, the more likely is the refinement to be valid. That is an example of Lemma 4, and is an important practical point: one need not examine *all* invariant declarations in order to apply a particular law.

Also, Law 4 illustrates the general coding process: it replaces a non-executable construct, the specification, with code. Similar laws for the remaining constructs (but without invariants) can be found in [5, 48, 50]; the development method itself is the subject of [46].

5 Laws for local invariants

Local invariants are introduced with this law, whose proof uses Lemma 5 following.

Law 5 *Local invariant.* For any I, J, and P,

$$\{J\}\, P\, [J] \quad \sqsubseteq_I \quad |[\, \mathbf{inv}\ J \bullet P\,]|.$$

Proof: Suppose $K \Rrightarrow I$. Then

$$wp_K(\{J\}\, P\, [J], \phi)$$

$$
\begin{aligned}
&\equiv && J \wedge K \wedge wp_K(P[J], \phi) \\
&\equiv && J \wedge wp_K(P, K \wedge (J \Rightarrow \phi)) \\
&\equiv && \text{"Lemma 2"} \\
&   && J \wedge wp_K(P, J \Rightarrow \phi) \\
&\Rightarrow && \text{"Lemma 5"} \\
&   && wp_{K \wedge J}(P, \phi) \\
&\equiv && wp_K(|[\ \textbf{inv}\ J \bullet P\ ]|, \phi).
\end{aligned}
$$

♡

Lemma 5 *Maintain invariant.*

$$J \wedge wp_I(P, J \Rightarrow \phi) \Rrightarrow wp_{I \wedge J}(P, \phi).$$

Proof: Structural induction over P; in fact equivalence $\equiv$ holds in every case except sequential composition.
♡

Note that in Law 5 the left-hand side assumes J before P and establishes it after P; the right-hand side maintains it within P as well.

Local invariants can also be introduced implicitly, within typed local variable declarations:

Law 6 *Introduce local block.* Provided x is a fresh local variable, not occurring in T or *post*,

$$w:[\mathit{post}] \quad \sqsubseteq_I \quad |[\ \textbf{var}\ x\colon T\ \textbf{and}\ J \bullet w, x\colon [\mathit{post}]\ ]|.$$

Proof: Let $K \Rrightarrow I$, and assume x is a fresh variable. We introduce the abbreviation Φ for the formula $x \in T \wedge J$, and proceed

$$
\begin{aligned}
&   && wp_K(w\colon [\mathit{post}], \phi) \\
&\equiv && K \wedge (\forall w \bullet K \wedge \mathit{post} \Rightarrow \phi) \\
&\equiv && \text{"x is fresh"} \\
&   && (\forall x \bullet K \wedge (\forall w, x \bullet K \wedge \mathit{post} \Rightarrow \phi)) \\
&\Rightarrow && (\forall x \bullet K \wedge (\forall x \bullet K \wedge \Phi \Rightarrow K \wedge \Phi \wedge (\forall w, x \bullet K \wedge \Phi \wedge \mathit{post} \Rightarrow \phi))) \\
&\equiv && (\forall x \bullet wp_K(x\colon [\Phi], wp_{K \wedge \Phi}(w, x\colon [\mathit{post}], \phi))) \\
&\equiv && \text{"Definition 6"} \\
&   && wp_K(|[\ \textbf{var}\ x\colon T\ \textbf{and}\ J \bullet w, x\colon [\mathit{post}]\ ]|, \phi).
\end{aligned}
$$

♡

Theorem 1 below justifies *stepwise refinement* in context: refining a part of a program in context I refines the whole program in that context:

Theorem 1 *Monotonicity.* Let $\mathcal{C}$ be a program scheme containing the program name p, and let $\mathcal{C}(X)$ be the result of replacing all occurrences of p in $\mathcal{C}$ by the program X. Then for any programs P, Q, if $P \sqsubseteq_I Q$, then also $\mathcal{C}(P) \sqsubseteq_I \mathcal{C}(Q)$.

Proof: Structural induction over $\mathcal{C}$. The only novel case is local invariants: suppose $P \sqsubseteq_I Q$. If $J \Rightarrow I$, then

$$\begin{aligned}
& wp_J(|[\,\mathbf{inv}\ K \bullet P\,]|, \phi) \\
\equiv\ & wp_{J \wedge K}(P, \phi) \\
\Rightarrow\ & \text{"Lemma 3 and assumption, since } J \wedge K \Rightarrow I\text{"} \\
& wp_{J \wedge K}(Q, \phi) \\
\equiv\ & wp_J(|[\,\mathbf{inv}\ K \bullet Q\,]|, \phi).
\end{aligned}$$

♡

Theorem 2 improves Theorem 1, and it is *the* reason for introducing a local invariant, since within its scope we can use the refinement relation $\sqsubseteq_{I \wedge J}$, easier to establish than $\sqsubseteq_I$.

Theorem 2 *Use local invariant.* Let $\mathcal{C}$ be as before. If $P \sqsubseteq_{I \wedge J} Q$, then

$$|[\,\mathbf{inv}\ J \bullet \mathcal{C}(P)\,]| \quad \sqsubseteq_I \quad |[\,\mathbf{inv}\ J \bullet \mathcal{C}(Q)\,]|.$$

Proof: From Theorem 1, $\mathcal{C}(P) \sqsubseteq_{I \wedge J} \mathcal{C}(Q)$. The result follows from Definitions 8 and 5.
♡

Note that the stronger hypothesis $P \sqsubseteq_I Q$ would be enough in Theorem 2: we may use the new invariant J, but are not obliged to do so.

Although local invariants make refinement easier, there is a price to pay later. After the refinement, still the **inv** remains — and it is not code. The next section deals with its elimination.

6 Eliminating local invariants

An implicit invariant, introduced by Law 6, need not be eliminated — because in that special case it *is* code. But explicit invariants (Law 5, or **and** in Law 6) must be removed, and that requires laws like these:

Law 7 *Invariant distribution through sequential composition.* For any context J,

$$|[\,\mathbf{inv}\ I \bullet P;\ Q\,]| \quad \sqsubseteq_J \quad |[\,\mathbf{inv}\ I \bullet P\,]|;\ |[\,\mathbf{inv}\ I \bullet Q\,]|.$$

Proof: Direct from Definition 5 and Figure 1. (In fact, the two programs are equal.)
♡

Law 8 *Invariant distribution through alternation.* For any context J, suppose that $J \wedge I \Rightarrow (G_i \Leftrightarrow G'_i)$ for each branch i of the alternation. Then

$$|[\,\mathbf{inv}\ I \bullet \mathbf{if}\ ([\!]i \bullet G_i \rightarrow P_i)\ \mathbf{fi}\,]|$$

$$\sqsubseteq_J \quad \mathbf{if}\ ([\!]i \bullet G'_i \rightarrow |[\,\mathbf{inv}\ I \bullet P_i]|\,)\ \mathbf{fi}.$$

Proof: Let $K \Rightarrow J$. Then

$$\begin{array}{rl}
& wp_K(lhs, \phi) \\
\equiv & (\bigvee i \bullet G_i) \wedge (\bigwedge i \bullet G_i \Rightarrow wp_{I \wedge K}(P_i, \phi)) \\
\Rightarrow & \text{``assumptions, Lemma 1''} \\
& (\bigvee i \bullet G'_i) \wedge (\bigwedge i \bullet G'_i \Rightarrow wp_{I \wedge K}(P_i, \phi)) \\
\equiv & wp_K(rhs, \phi).
\end{array}$$

♡

Law 9 *Invariant elimination for assignment.* For any context J, provided $J \wedge I \Rightarrow I[w \backslash E]$,

$$|[\,\mathbf{inv}\ I \bullet w := E\,]| \quad \sqsubseteq_J \quad w := E.$$

Proof: Let $K \Rightarrow J$. Then

$$\begin{array}{rl}
& wp_K(|[\,\mathbf{inv}\ I \bullet w := E\,]|, \phi) \\
\equiv & wp_{I \wedge K}(w := E, \phi) \\
\equiv & I \wedge K \wedge (I \wedge K \Rightarrow \phi)[w \backslash E] \\
\Rightarrow & \text{``assumptions''} \\
& K \wedge (K \Rightarrow \phi)[w \backslash E] \\
\equiv & wp_K(w := E, \phi).
\end{array}$$

♡

Law 10 *Invariant elimination for* **skip**. For any context J,

$$|[\, \text{inv } I \bullet \text{skip} \,]| \quad \sqsubseteq_J \quad \text{skip}.$$

Proof: Trivial.
♡

We see in Section 8.2 a law that distributes **inv** over recursion; and there are analogs of Laws 9 and 10 that deal with **abort**, annotations, and specifications. All of those, together, allow **inv** to be eliminated by first distributing it towards the atomic statements of a program, then discharging certain proof obligations at each separately. That last step, of which Law 9 is an example, is like type-checking, to which finally we turn.

7 Type-checking

Type-checking is the automated application, say by a complier, of the procedure described in Section 6 — but in the special case of implicit invariants for typed local variables.

If the types can be made only in certain ways (for example, enumeration, Cartesian product, disjoint union, etc.), and the expressions E in assignments are restricted similarly, then it is decidable whether an expression E is of type T given that we know the types of the constituents of E. Consider, for example, the typing context

$$a, b, c \in \mathbf{Z}, \tag{3}$$

and the assignment $a := b + c$. For the elimination of the invariant (3), we require by Law 9

$$a, b, c \in \mathbf{Z} \Rightarrow b + c, b, c \in \mathbf{Z}.$$

And that follows from $b, c \in \mathbf{Z} \Rightarrow b + c \in \mathbf{Z}$, which can be built in to a compiler.

But now consider a more interesting case: let the invariant be $I \;\hat{=}\; m, n \in \mathbf{N}$. By Law 4 (taking *pre* to be *true*), we have

$$n \colon [n = m - 1] \quad \sqsubseteq_I \quad n := m - 1.$$

And so by Definition 6 and Theorems 2, 1, we have

$$|[\ \textbf{var}\ m, n : \mathbb{N} \bullet n\colon [n = m - 1]\ ]|$$
$$\sqsubseteq\ |[\ \textbf{var}\ m, n : \mathbb{N} \bullet n := m - 1\ ]|.$$

To the experienced programmer, that looks unlikely: if the value of m is zero, surely the assignment will abort — yet the specification does not abort! And we are not saved either by (decidable) type checking, unless *all* such assignments are ill-typed: a type checker cannot know the actual value of m.

In fact, all such assignments *are* ill-typed. Natural number subtraction is not integer subtraction: it differs exactly in the case where the result would be negative. Using $\ominus$ for natural number subtraction, the assignment $n := m \ominus 1$ is well typed, but the earlier refinement fails:

$$n\colon [n = m - 1]\quad \not\sqsubseteq_I \quad n := m \ominus 1.$$

And that is where it should fail.

Suppose, then, that we know m is positive. In that case we must show

$$\{m > 0\} n\colon [n = m - 1]\quad \sqsubseteq_I \quad n := m \ominus 1,$$

and that follows from Law 4 provided

$$I \wedge m > 0 \Rightarrow m - 1 = m \ominus 1.$$

The proviso is clearly true.

Fortunately, most operators can still be overloaded; we needn't distinguish natural and integer addition, for example. And some of the distinctions are already in widespread use: compare / and **div**.

Note further that Definition 6 introduces an initialising specification. That can be refined to an assignment, provided the type is non-empty — which, therefore, we require.

Thus, within the above constraints, we can view type-checking as the elimination of a specific kind of local invariant, and it can be done automatically by a compiler.

8 Recursion

8.1 Syntax and semantics

A recursive program is written

$$\mathbf{mu}\ p \bullet \mathcal{C}\ \mathbf{um}, \tag{4}$$

where p is a program name and $\mathcal{C}$ a program scheme probably containing p. For its meaning, we must understand $\mathcal{C}$ as a function from programs to programs: the application of $\mathcal{C}$ to the program X is just $\mathcal{C}(X)$, the program left when p is replaced in $\mathcal{C}$ by X. The meaning of (4) is then the least fixed-point of that function.

It is not the meanings of programs that must be monotonic for such least fixed points to exist; and indeed those meanings are not monotonic in their context argument. For example, from Figure 1 we have

$$\begin{aligned} wp_{false}(x := 0, x \neq 0) &\equiv false \\ wp_{x \neq 0}(x := 0, x \neq 0) &\equiv true \\ wp_{true}(x := 0, x \neq 0) &\equiv false. \end{aligned}$$

It is the program constructors that must be monotonic over programs; and their monotonicity is stated in Theorem 1. (The monotonicity of $\mathbf{mu}\cdots\mathbf{um}$ follows from the monotonicity of *fix*, the least fixed-point operator itself.)

8.2 Eliminating local invariants

We must extend Section 6 with a law that allows **inv** to be eliminated when it surrounds a recursion. This is the law:

Law 11 *Invariant elimination for recursion.* Let $\mathcal{C}$ and $\mathcal{D}$ be two program schemes, and for any program X let $\mathcal{C}(X)$ and $\mathcal{D}(X)$ be the programs resulting when the program name p is replaced by X. Suppose that for any program X,

$$|[\ \mathbf{inv}\ I \bullet \mathcal{C}(X)\ ]| \ \sqsubseteq_J \ \mathcal{D}(|[\ \mathbf{inv}\ I \bullet X\ ]|).$$

Then we may eliminate a surrounding invariant as follows:

$$|[\ \mathbf{inv}\ I \bullet \mathbf{mu}\ p \bullet \mathcal{C}\ \mathbf{um}\ ]| \ \sqsubseteq_J \ \mathbf{mu}\ p \bullet \mathcal{D}\ \mathbf{um}.$$

Proof: Suppose $K \Rightarrow J$. Then

$$\begin{align}
&\quad wp_K(\|[\text{ inv } I \bullet \text{mu } p \bullet \mathcal{C} \text{ um }]\|, \phi)\\
\equiv &\quad wp_{K \wedge I}(\text{mu } p \bullet \mathcal{C} \text{ um}, \phi)\\
\equiv &\quad \text{fix } \mathcal{C} \ (K \wedge I) \ \phi,
\end{align}$$

where we use $\mathcal{C}$ also for the *meaning* of the program scheme. (A more exact treatment would use semantic functions with environments, in the style of [61].) Since $\mathcal{C}$ is monotonic, we can continue

$$\begin{align}
\equiv &\quad \text{"for some ordinal } \beta\text{"}\\
&\quad (\bigvee_{\alpha<\beta} \mathcal{C}^\alpha \text{ abort } (K \wedge I) \ \phi)\\
\equiv &\quad (\bigvee_{\alpha<\beta} \|[\text{ inv } I \bullet \mathcal{C}^\alpha \text{ abort }]\| \ K \ \phi)\\
\Rightarrow &\quad \text{"assumption, and transfinite induction over } \alpha\text{"}\\
&\quad (\bigvee_{\alpha<\beta} \mathcal{D}^\alpha \ \|[\text{ inv } I \bullet \text{ abort }]\| \ K \ \phi)\\
\equiv &\quad (\bigvee_{\alpha<\beta} \mathcal{D}^\alpha \text{ abort } K \ \phi)\\
\Rightarrow &\quad wp_K(\text{mu } p \bullet \mathcal{D} \text{ um}, \phi).
\end{align}$$

♡

Given a program scheme $\mathcal{C}$, in practice the $\mathcal{D}$ required by Law 11 is found as before: $\|[\text{ inv} \cdots]\|$ is distributed inwards until it reaches the recursive call p; then it is removed. What results is $\mathcal{D}$. An example is given in the next section.

8.3 Iteration

Iteration, written do $([]i \bullet G_i \to P_i)$ od, is just an abbreviation for this recursion:

$$\begin{align}
&\text{mu } p \bullet\\
&\quad \text{if } ([]i \bullet G_i \to P_i; p)\\
&\quad [] \ \neg(\forall i \bullet G_i) \to \text{skip}\\
&\quad \text{fi}\\
&\text{um.}
\end{align}$$

That completes Figure 1 for Dijkstra's original language. Now by Laws 8,

7, 10, we have for any context J and program X that the program

$$|[\text{ inv } I \bullet$$
$$\quad \text{if } ([\!]i \bullet G_i \to P_i; X)$$
$$\quad [\!] \neg (\vee i \bullet G_i) \to \text{skip}$$
$$\quad \text{fi}$$
$$]|$$

is refined under $\sqsubseteq_J$ by

$$\text{if } ([\!]i \bullet G'_i \to |[\text{ inv } I \bullet P_i]|; |[\text{ inv } I \bullet X]|)$$
$$[\!] \neg (\vee i \bullet G'_i) \to \text{skip}$$
$$\text{fi},$$

provided that, for each i, $J \wedge I \Rrightarrow (G_i \Leftrightarrow G'_i)$. Law 11 then gives immediately

Law 12 *Invariant distribution through iteration.* Providing, for each i, that $J \wedge I \Rrightarrow (G_i \Leftrightarrow G'_i)$,

$$\sqsubseteq_J \quad \begin{array}{l} |[\text{ inv } I \bullet \text{do } ([\!]i \bullet G_i \to P_i) \text{ od}]| \\ \text{do } ([\!]i \bullet G'_i \to |[\text{ inv } I \bullet P_i]|) \text{ od} \end{array}$$

♡

9 Examples

Our examples are chosen to expose the difference in practice between explicit and implicit invariants: we present two developments of a program to calculate the greatest common divisor.

In the first example we use an explicit invariant; since it is not code, it must be removed "by hand" at the very end. During the development, however, it provides extra context. In the second example we use an implicit invariant instead, and during development we respect the type constraint it imposes. Then there is nothing significant to remove at the end.

Both developments require laws of refinement not already presented, and those have been placed in the appendix. Also required are logical constants and initial variables, which we now explain.

9.1 Logical constants and initial variables

Consider the following specification that x must increase:

$$\{X = x\} \; x\colon [x > X] \,. \tag{5}$$

It refines to $x := x+1$ for example — but it refines also to $x := X+1$. Usually the second refinement is not intended, since X is only a place-holder for the initial value of x and should not appear in the final program.

In fact X above is a *logical constant*, which we now define precisely [46]. Logical constants are declared using **con**, and their scope indicated by a local block. This is the definition:

Definition 9 *Logical constant.* Provided neither I nor ϕ contain free X,

$$wp_I(|[\, \mathbf{con}\; X \bullet P \,]|, \phi) \;\;\widehat{=}\;\; (\exists X \bullet wp_I(P, \phi)) \,.$$

♡

Logical constants need not be in upper case, of course, though for clarity we will follow that convention.

Definition 9 allows Specification (5) to be rewritten

$$|[\, \mathbf{con}\; X \bullet \{X = x\} \; x\colon [x > X] \,]|, \tag{6}$$

and it can no longer be refined to $x := X + 1$.

Following are laws for introducing and removing logical constants.

Law 13 *Introduce logical constant.*

$$P \;\;\sqsubseteq_I\;\; |[\, \mathbf{con}\; X \bullet P \,]| \,.$$

♡

Note that Law 13 applies whether or not X is free in P; usually, however, the introduced logical constant is a fresh name.

Since a logical constant is not code, it must be removed after it has served its purpose; logical constants are used during development, but are not executed. They can be removed when all references to them have been eliminated:

Law 14 *Remove logical constant.* If X occurs nowhere in program P, then

$$|[\ \mathbf{con}\ X \bullet P\ ]| \sqsubseteq_I P.$$

♡

There are many uses of **con**, and we shall see some in our examples to follow. But the most common use is to refer to initial values, and we introduce an abbreviation especially for that:

Abbreviation 1 *Initial variables.* Occurrences of 0-subscripted variables in the postcondition of a specification refer to values held by those variables in the *initial* state. Let x be any variable, probably occurring in the frame w; if X is a fresh name, then

$$\{pre\}\ w \colon [post] \quad \hat{=} \quad |[\ \mathbf{con}\ X \bullet \{pre \wedge X = x\}\ w \colon [post[x_0 \backslash X]]\ ]|.$$

We reserve 0-subscripted names for that purpose, and call then *initial variables*.

♡

Using Abbreviation 1, Specification (6) is written $x \colon [x > x_0]$, since trivially we have $\{true\} = \mathbf{skip}$.

A more detailed discussion of logical constants and initial variables appears in [46].

9.2 First example: explicit invariant

In the first example, we assume the two numbers whose gcd is to be found are both integers: that is, the development is carried out in the context of a declaration **var** $a, b \colon \mathbb{Z}$. The algorithm we derive assumes as well, however, that both are non-negative initially, as reflected in its specification:

$$|[\ \mathbf{con}\ G \bullet \\ \{a, b \geq 0 \wedge G = \gcd(a, b)\}\ a, b \colon [a = G] \\ ]|. \quad \triangleleft$$

The sign ◁ in the right margin indicates the part of the program we next refine; the surrounding text is unchanged.

We decide that we will maintain the condition $a, b \geq 0$ throughout, and by making it an invariant we avoid having to carry it explicitly through the development. We proceed (in small steps for illustration)

⊑ "Law 15 *Split assumption*"

$\{a, b \geq 0\} \; \{G = \gcd(a, b)\} \; a, b \colon [a = G]$

⊑ "Law 3 *Annotations and* **skip**"

$\{a, b \geq 0\} \; \{G = \gcd(a, b)\} \; a, b \colon [a = G] \, ; \; [a, b \geq 0]$

⊑ "Law 5 *Local invariant*"

|[**inv** $a, b \geq 0$ •
    $\{G = \gcd(a, b)\} \; a, b \colon [a = G]$ ◁
]|.

Note that all the refinements above occur in the context $a, b \in \mathbb{Z}$ provided by the assumed declaration of a and b as integers. Thus the relation between successive lines is actually $\sqsubseteq_{a, b \in \mathbb{Z}}$, though that would be tedious to write out each time. We assume therefore in setting out developments that the refinements are relative to all enclosing invariants.

Now we anticipate an iteration terminating with one of a, b equal to zero, and the other holding the *gcd*. So we make the following step, again refining only the part indicated by ◁ above:

⊑ "Law 16 *Following assignment*"

$\{G = \gcd(a, b)\} \; a, b \colon [a + b = G] \, ;$ ◁
$a \coloneq a + b.$

The refinement relation this time (and from here on) is $\sqsubseteq_{a, b \in \mathbb{N}}$, since we are now enclosed by the invariants $a, b \in \mathbb{Z}$ (implicit) and $a, b \geq 0$ (explicit).

The iteration is then introduced using Law 17; the invariant and variant are included as comments in the development. Formulae stacked vertically are implicitly conjoined.

⊑ "invariant: $G = \gcd(a, b)$; variant: $a + b$"

do $a \geq b > 0 \rightarrow$
    $\left\{ \begin{array}{l} a \geq b > 0 \\ G = \gcd(a, b) \end{array} \right\} \; a, b \colon \left[\begin{array}{l} a+b < a_0+b_0 \\ G = \gcd(a, b) \end{array} \right]$ (i)

[] $b \geq a > 0 \rightarrow$
    $\left\{ \begin{array}{l} b \geq a > 0 \\ G = \gcd(a, b) \end{array} \right\} \; a, b \colon \left[\begin{array}{l} a+b < a_0+b_0 \\ G = \gcd(a, b) \end{array} \right]$ (ii)

od.

Although Law 17 bounds the variant below by 0, we leave that out in (i) and (ii) since our context provides it. (More precisely, we use Law 2 to remove it.)

Now the two conditions required by Law 17 for the above step are

- $a, b \in \mathbb{N} \land G = \gcd(a, b) \Rightarrow G = \gcd(a, b)$

- $a, b \in \mathbb{N} \land \left(\begin{array}{c} G = \gcd(a, b) \\ (a < b \lor b \leq 0) \\ (b < a \lor a \leq 0) \end{array} \right) \Rightarrow a + b = G.$

The first condition is obviously met, and the second follows from this:

$$a, b \in \mathbb{N} \land \left(\begin{array}{c} G = \gcd(a, b) \\ (a = 0 \lor b = 0) \end{array} \right) \Rightarrow a + b = G.$$

Finally, we complete the development by refining the guarded statements to assignments as follows:

(i) $\sqsubseteq$ "Law 4 *Assignment*"

   $a := a - b$

(ii) $\sqsubseteq$ "Law 4 *Assignment*"

   $b := b - a.$

We leave out the assumptions (Law 3), and in the proviso assume additionally that $a = a_0 \land b = b_0$. Strictly speaking, that is an extension of Law 4 to account for Abbreviation 1.

At this point the collected program is

```
|[ con G; inv a, b ≥ 0 •
    do  a ≥ b > 0 → a := a - b
    []  b ≥ a > 0 → b := b - a
    od;
    a := a + b
]|.
```

Except for the declarations **con** and **inv**, we have reached code. And since G is not used in the program, the **con** is removed by Law 14.

The invariant $a, b \geq 0$ is not so easy to remove. We distribute it inwards, using Laws 7 and 12. That gives

$$\begin{array}{l} \textbf{do} \ \ a \geq b > 0 \rightarrow |[\ \textbf{inv}\ a, b \geq 0 \bullet a := a - b\ ]| \\ [\!]\ \ \ \ b \geq a > 0 \rightarrow |[\ \textbf{inv}\ a, b \geq 0 \bullet b := b - a\ ]| \\ \textbf{od}; \\ |[\ \textbf{inv}\ a, b \geq 0 \bullet a := a + b\ ]|. \end{array}$$

Of the three inner invariant blocks, only the last can be immediately replaced by its body (using Law 9), because the associated condition for that removal is trivially true:

$$a, b \in \mathbb{N} \ \ \Rightarrow \ \ a+b, b \geq 0.$$

But to use that law for the other invariant blocks requires the conditions

$$\begin{array}{l} a, b \in \mathbb{N} \ \ \Rightarrow \ \ a-b, b \geq 0 \\ a, b \in \mathbb{N} \ \ \Rightarrow \ \ a, b-a \geq 0, \end{array}$$

neither of which is true. In fact, elimination of those invariants requires an assumption which we have discarded — if in using Law 17 we had retained (and weakened by Law 1) the loop guards as assumptions, we would have instead

$$\begin{array}{l} \textbf{do} \ \ a \geq b > 0 \rightarrow |[\ \textbf{inv}\ a, b \geq 0 \bullet \{a \geq b > 0\}\ a := a - b\ ]| \\ [\!]\ \ \ \ b \geq a > 0 \rightarrow |[\ \textbf{inv}\ a, b \geq 0 \bullet \{b \geq a > 0\}\ b := b - a\ ]| \\ \textbf{od}; \\ a := a + b. \end{array}$$

Now the inner invariant blocks can be eliminated using Law 18, giving the following program:

$$\begin{array}{l} \textbf{do} \ \ a \geq b > 0 \rightarrow a := a - b \\ [\!]\ \ \ \ b \geq a > 0 \rightarrow b := b - a \\ \textbf{od}; \\ a := a + b. \end{array}$$

The automatic type-checking finally required by the original declaration **var** $a, b : \mathbb{Z}$ is trivial, since $\mathbb{Z}$ is closed under both addition and subtraction. That completes the development.

Notice that the program fragment (i) could have been refined even to the ridiculous $a := a - 5b$, say, but would never then have led to code. In removing the invariant, we would have reached

$$|[\ \textbf{inv}\ a, b \geq 0 \bullet \{a \geq b > 0\}\ a := a - 5b\ ]|.$$

And there we would have remained, as we cannot satisfy the proviso

$$a, b \in \mathbb{N} \land a \geq b > 0 \quad \Rightarrow \quad a-5b, b \geq 0.$$

9.3 Second example: implicit invariant

In the second example, we assume the two numbers whose *gcd* is to be found are both natural numbers: that is, that the development is carried out in the context of a more restrictive declaration **var** $a, b: \mathbb{N}$. There is no need to record an explicit assumption that a and b are non-negative, therefore, since that is known from their type. We begin with

$$|[\ \text{con}\,G \bullet$$
$$\{G = \gcd(a,b)\}\ a, b:\ [a = G]$$
$$]|.$$
◁

As before, we introduce the final assignment and the loop.

$$\sqsubseteq\ \{G = \gcd(a,b)\}\ a, b:\ [a + b = G]\,;$$
$$a := a + b$$
◁

$\sqsubseteq$ "invariant: $G = \gcd(a,b)$; variant: $a + b$"

$$\textbf{do}\ a \geq b > 0 \to$$
$$\left\{\begin{array}{l} a \geq b > 0 \\ G = \gcd(a,b) \end{array}\right\}\ a, b:\ \left[\begin{array}{l} a+b < a_0+b_0 \\ G = \gcd(a,b) \end{array}\right] \qquad (i)$$
$$[\!]\ \ b \geq a > 0 \to$$
$$\left\{\begin{array}{l} b \geq a > 0 \\ G = \gcd(a,b) \end{array}\right\}\ a, b:\ \left[\begin{array}{l} a+b < a_0+b_0 \\ G = \gcd(a,b) \end{array}\right] \qquad (ii)$$
$$\textbf{od}.$$

Note the simplification again of $0 \leq a+b < a_0+b_0$, this time justified because it is in the scope of the implicit invariant $a, b \in \mathbb{N}$.

The guarded statements are refined by assignments, but we use natural number subtraction so that the subsequent type-checking will succeed:

(i) $\sqsubseteq\ a := a \ominus b$

(ii) $\sqsubseteq\ b := b \ominus a$.

It would have been correct at this point to use ordinary subtraction $a - b$ as before, but that would fail the type checking later imposed by **var** $a, b: \mathbb{N}$. It would have been correct, too, in the first example to use $a \ominus b$; then the removal of the explicit invariant would not have required the re-introduced assumptions.

Collecting the program at this point reveals

```
|[ con G •
    do  a ≥ b > 0 → a:= a ⊖ b
    []  b ≥ a > 0 → b:= b ⊖ a
    od;
    a:= a + b
]|.
```

The logical constant G is removed as before, leaving the final program

```
do  a ≥ b > 0 → a:= a ⊖ b
[]  b ≥ a > 0 → b:= b ⊖ a
od;
a:= a + b.
```

Now the type-checking (described in Section 7) of the above code can be performed automatically, and succeeds since **N** is closed under $\ominus$. But **N** is not closed under ordinary subtraction, so using $a - b$ would have caused type checking to fail.

Our two examples show that an implicit invariant is more convenient than the equivalent explicit one, since then nothing need be explicitly removed. The type checking is automatic, and that corresponds to ordinary programming practice. Explicit invariants, however, allow the rigorous use of invariants that cannot be decidably type-checked. The benefit is that they, like types, can be assumed everywhere and need not be copied from place to place; the price is the explicit reasoning, finally, to remove them.

10 A discussion of motives

The definition of $wp_I(\sqcup, \sqcup)$ was suggested by a certain kind of data refinement. Let $P \leq P'$ mean that P is data-refined to P' under the transformation

(no abstract variables, no concrete variables, coupling invariant I).

Such data refinements are described in [41]. From the definitions there, we have that if $P \leq P'$ then for all ϕ,

$$I \wedge wp(P, \phi) \Rightarrow wp(P', I \wedge \phi).$$

It can be shown that the least-refined such P' is defined by

$$wp(P', \phi) \quad = \quad I \wedge wp(P, I \Rightarrow \phi), \tag{7}$$

and that is the motivation for Lemma 5. (Incidentally, there is a corresponding formula for data refinements in general: it is

$$(\exists a \bullet I \wedge wp(P, (\forall c \bullet I \Rightarrow \phi))).$$

We have just taken the special case in which the lists a and c of abstract and concrete variables are both empty.)

Equation (7) is where the definitions of Figure 1 and Section 3.1 come from, and it is the uniform application of that which gives an invariant-breaking assignment its miraculous semantics, just as ill-advised data refinements lead to miracles [42].

We know that data refinement has nice distribution properties, and that is why $|[\,\mathbf{inv}\, I \bullet \cdots \,]|$ does too. One can see

$$|[\,\mathbf{inv}\, I \bullet P \,]|$$

as the program got by distributing the data refinement, above, through P. And that is why Lemma 5 is true: such distribution can only refine the program.

11 Related work

The refinement calculus, first proposed by Back [5], has in fact been invented twice more [44, 50]. At Oxford it was made specifically for the rigorous development of programs from Z specifications [25, 60].

A prominent feature of Z specifications is the *schema* which, when used to describe abstract operations, carries an invariant around with it that includes type information and is maintained automatically. That is where we started.

Where we have finished is very similar to work by Lamport and Schneider [36]. Their **constraints** clause and our **and** declaration have the same effect; their *Constraint strengthening rule* is like our Law 5. Lamport and Schneider use *partial* correctness, however, and do not write specifications within their programs. Their constructions are defined within temporal logic; ours are defined by weakest preconditions.

Lamport and Schneider's use of partial correctness identifies aborting and miraculous behaviour, leading them to say that invariant-breaking assignments abort. For them, such programs establish *false* if they terminate — therefore they don't terminate. Ours "damn the torpedoes" and terminate anyway. Section 12 explains why.

Within our refinement calculus, Lamport and Schneider's constraint $x \perp y$ — "x and y are independent" — would be expressed instead as a dependency (their **may alias**). That dependency would in Definition 1 link the variables in the frame to the bound variables in its meaning, allowing a more general relationship than our present equality. The frame would be expanded, according to the aliasing dependencies, before being applied as a universal quantification. Finally, their proposed extension to generalised assignments $exp := exp'$ is already neatly done with specifications.

Invariants are used also by Reynolds [57], called *general invariants*, and are true from their point of occurrence to the end of the smallest enclosing block. But the *specification logic* does not give them a meaning, nor are they connected with type information. The temporary falsification of a general invariant is allowed, however, and we have not discussed that here: there are several approaches to pursue. Like Lamport and Schneider, Reynolds uses partial correctness.

12 Conclusions

Although the traditional $wp(\sqcup, \sqcup)$ is our $wp_{true}(\sqcup, \sqcup)$, the traditional $\sqsubseteq$, as in [5, 27, 44, 50], is *not* our $\sqsubseteq$. That is because Definition 7 insists that the implication hold for all contexts.

What have we lost? Not much. Here is an example; with our definition of refinement,

$$n := -1; \; n := 0 \quad \not\sqsubseteq \quad n := 0.$$

Just take the context $n \in \mathbb{N}$, and the left-hand side becomes miraculous: it cannot refine to code. But most refinement laws have *atomic* left-hand sides — after all, they are there to introduce structure, not remove it. And when the left-hand side is atomic, we do preserve the traditional refinement relation — because that is true for any data-refinement [20, 51, 5]. Thus most existing refinement laws remain valid. For example (the left-hand side is atomic), we still have

$$n := 0 \sqsubseteq n := -1; \; n := 0.$$

But the price of the miraculous assignment, when it appears to break the invariant, is the final type-checking. The "obvious" alternative is the operationally-motivated alternative definition

$$wp_I(x := E, \phi) \quad \hat{=} \quad I \wedge (I \wedge \phi)[x \backslash E].$$

All that does, however, is add the type-checking to Law 4, and we lose, *formally* at least, the ability to delay such checking until the final test for feasibility. More significant, however, is that using the above definition would not allow the refinement

$$x\colon [x = E] \quad \sqsubseteq \quad x := E.$$

(We assume that E contains no x.) For if that refinement were valid, then by Lemma 4 it would be valid in the context $x \neq E$ as well. But it can't be: in that context, the left-hand side is a miracle but, with the alternative definition of assignment, the right-hand side would abort.

The ability to factor "details" like feasibility, and hence type-checking, is essential in a practical method. Experienced programmers will only be impeded by continual checking of types: their programs tend to be well-typed anyway. And in cases of error, the compiler acts at the last minute, catching the mistake. Because they are experienced, that will happen rarely.

Inexperienced programmers, however, will waste a lot of time by leaving their feasibility and type checks till later. When eventually the check is made, and fails, all the intervening development must be discarded. Because they are inexperienced, that will happen often.

But we cannot exploit experience by allowing those having it to apply "only some of the rules". All programmers, experienced or not, must apply all of the rules; but for each, the set of rules to which "all" applies may be different. The experts' rules are those from which all feasibility and type checks have been removed; the rules for apprentices have all the checks built in, and performance suffers. A mathematical factorisation is necessary to make that distinction reliably; the one we have chosen is close to what people do already.

Acknowledgement

We thank Paul Gardiner for his help, and Ian Hayes for his comments on the earlier version [45].

A Additional refinement laws

These laws are used in Section 9.

Law 15 *Split assumption.*

$$\{pre \wedge pre'\} \quad \sqsubseteq_I \quad \{pre\}\,\{pre'\}.$$

♡

Law 16 *Following assignment.* For any term E,

$$w, x\colon [post]$$
$$\sqsubseteq_I \quad w, x\colon [post[x\backslash E]]; \\ x := E.$$

♡

Law 17 *Loop introduction.* Providing $I \wedge pre \Rightarrow inv$ and $I \wedge inv \wedge \neg(\bigvee G_i) \Rightarrow post$, and v is any integer-valued expression,

$$\{pre\}\ w\colon [post]$$
$$\sqsubseteq_I \quad \mathbf{do}\ (\mathord{\|}\ i \bullet G_i \to \{G_i \wedge inv\}\ w\colon [inv \wedge 0 \leq v < v_0])\ \mathbf{od}.$$

♡

Law 18 *Invariant elimination from preconditioned assignment.* Provided $J \wedge I \wedge pre \Rightarrow I[w\backslash E]$,

$$|[\ \mathbf{inv}\ I \bullet \{pre\}\ w := E\ ]| \quad \sqsubseteq_J \quad w := E.$$

♡

References

[1] M. Abadi and L. Lamport. The existence of refinement mappings. Technical Report 29, Digital Systems Research Center, August 1988.

[2] J.-R. Abrial. Generalised substitutions. 26 Rue des Plantes, Paris 75014, France, 1987.

[3] J.-R. Abrial. A formal approach to large software construction. In J.L.A. van de Snepscheut, editor, *Mathematics of Program Construction*, volume 375 of *Lecture Notes in Computer Science*, pages 1-20. Springer, June 1989.

[4] R.-J.R. Back. On the correctness of refinement steps in program development. Report A-1978-4, Department of Computer Science, University of Helsinki, 1978.

[5] R.-J.R. Back. Correctness preserving program refinements: Proof theory and applications. Tract 131, Mathematisch Centrum, Amsterdam, 1980.

[6] R.-J.R. Back. Procedural abstraction in the refinement calculus. Report Ser.A 55, Departments of Information Processing and Mathematics, Swedish University of Åbo, Åbo, Finland, 1987.

[7] R.-J.R. Back. A calculus of refinements for program derivations. *Acta Informatica*, 25:593-624, 1988.

[8] R.-J.R. Back. Data refinement in the refinement calculus. In *Proceedings 22nd Hawaii International Conference of System Sciences*, Kailua-Kona, January 1989.

[9] R.-J.R. Back. Refinement calculus ii: Parallel and reactive programs. In J.W. deBakker, W.P. deRoever, and G. Rozenberg, editors, *Lecture Notes in Computer Science 430*, pages 67-93. Springer, 1990.

[10] R.-J.R. Back and K. Sere. Stepwise refinement of parallel algorithms. *Science of Computer Programming*, 13(2-3):133-180, 1990.

[11] R.J.R. Back and J. von Wright. Statement inversion and strongest postcondition. Reports on Computer Science and Mathematics, Åbo Akademi, 1989.

[12] H. Barringer, J.H. Cheng, and C.B. Jones. A logic covering undefinedness in program proofs. *Acta Informatica*, 21:251-269, 1984.

[13] F.L. Bauer, M. Broy, R. Gnatz, W. Hesse, and B. Krieg-Brückner. A wide spectrum language for program development. In *3rd International Symposium on Programming*, pages 1-15, Paris, 1978.

[14] R.S. Bird. An introduction to the theory of lists. Technical monograph PRG-56, Programming Research Group, 8-11 Keble Road, Oxford OX1 3QD, U.K., October 1986.

[15] H. Boom. A weaker precondition for loops. *ACM Transactions on Programming Languages and Systems*, 4:668–677, 1982.

[16] Wei Chen and J.T. Udding. Towards a calculus of data refinement. In J.L.A. van de Snepsheut, editor, *Lecture Notes in Computer Science 375: Mathematics of Program Construction*. Springer, June 1989.

[17] E.W. Dijkstra. *A Discipline of Programming*. Prentice-Hall, Englewood Cliffs, 1976.

[18] P.H.B. Gardiner. Data refinement of maps. In draft, August 1990.

[19] P.H.B. Gardiner and C.C Morgan. A single complete rule for data refinement. Technical Report PRG–TR–7–89, Programming Research Group, Oxford University, November 1989. In this collection.

[20] P.H.B. Gardiner and C.C. Morgan. Data refinement of predicate transformers. *Theoretical Computer Science*, 87:143–162, 1991. In this collection.

[21] D. Gries. *The Science of Programming*. Springer, 1981.

[22] D. Gries and D. Levin. Assignment and procedure call proof rules. *ACM Transactions on Programming Languages and Systems*, 2(4), October 1980.

[23] D. Gries and J. Prins. A new notion of encapsulation. In *Symposium on Language Issues in Programming Environments*. SIGPLAN, June 1985.

[24] J.V. Guttag, J.J. Horning, and J.M. Wing. Larch in five easy pieces. Technical Report 5, Digital Systems Research Center, July 1985.

[25] I.J. Hayes. *Specification Case Studies*. Prentice-Hall, London, 1987.

[26] J.F. He, C.A.R. Hoare, and J.W. Sanders. Data refinement refined. In *Lecture Notes in Computer Science 213*, pages 187–196. Springer, 1986.

[27] E.C.R. Hehner. *The Logic of Programming*. Prentice-Hall, London, 1984.

[28] C.A.R. Hoare. An axiomatic basis for computer programming. *Communications of the ACM*, 12(10):576–580, 583, October 1969.

[29] C.A.R. Hoare. Procedures and parameters: An axiomatic approach. In E. Engeler, editor, *Lecture Notes in Mathematics 188*. Springer, 1971.

[30] C.A.R. Hoare. Proof of correctness of data representations. *Acta Informatica*, 1:271–281, 1972.

[31] C.A.R. Hoare and J.F. He. The weakest prespecification. *Fundamenta Informaticae*, IX:51–84, 1986.

[32] C.A.R. Hoare, J.F. He, and J.W. Sanders. Prespecification in data refinement. *Information Processing Letters*, 25(2), May 1987.

[33] C.B. Jones. *Systematic Software Development using VDM*. Prentice-Hall, 1986.

[34] M.B. Josephs. Formal methods for stepwise refinement in the Z specification language. Programming Research Group, Oxford.

[35] M.B. Josephs. The data refinement calculator for Z specifications. *Information Processing Letters*, 27:29–33, 1988.

[36] L. Lamport and F.B. Schneider. Constraints: a uniform approach to aliasing and typing. In *12th Symposium on Principles of Programming Languages*, pages 205–216, New Orleans, January 1985. ACM.

[37] P. Lucas. Two constructive realizations of the block concept and their equivalence. Technical Report TR 25.085, IBM Laboratory Vienna, 1968.

[38] C.E. Martin, P.H.B. Gardiner, and O. de Moor. An algebraic construction of predicate transformers. Submitted for publication, 1991.

[39] L. Meertens. Abstracto 84: The next generation. In *Annual Conference*. ACM, 1979.

[40] C.C. Morgan. Software engineering course notes. In draft.

[41] C.C. Morgan. Auxiliary variables in data refinement. *Information Processing Letters*, 29(6):293–296, December 1988. In this collection.

[42] C.C. Morgan. Data refinement using miracles. *Information Processing Letters*, 26(5):243–246, January 1988. In this collection.

[43] C.C. Morgan. Procedures, parameters, and abstraction: Separate concerns. *Science of Computer Programming*, 11(1):17–28, 1988. In this collection.

[44] C.C. Morgan. The specification statement. *ACM Transactions on Programming Languages and Systems*, 10(3), July 1988. In this collection.

[45] C.C. Morgan. Types and invariants in the refinement calculus. In J.L.A. van de Snepsheut, editor, *Lecture Notes in Computer Science 375: Mathematics of Program Construction*. Springer, June 1989. In this collection.

[46] C.C. Morgan. *Programming from Specifications*. Prentice-Hall, 1990.

[47] C.C. Morgan and P.H.B. Gardiner. Data refinement by calculation. *Acta Informatica*, 27:481–503, 1990. In this collection.

[48] C.C. Morgan and K.A. Robinson. Specification statements and refinement. *IBM Journal of Research and Development*, 31(5), September 1987. In this collection.

[49] C.C. Morgan and B.A Sufrin. Specification of the UNIX filing system. *IEEE Transactions on Software Engineering*, SE–10(2), March 1984.

[50] J.M. Morris. A theoretical basis for stepwise refinement and the programming calculus. *Science of Computer Programming*, 9(3):287–306, December 1987.

[51] J.M. Morris. Laws of data refinement. *Acta Informatica*, 26:287–308, 1989.

[52] J.M. Morris. Nondeterministic expressions and the axiom of assignment. Unpublished, 1991.

[53] P. Naur (Ed.). Revised report on the algorithmic language Algol 60. *Communications of the ACM*, 6(1):1–17, January 1963.

[54] G. Nelson. A generalization of Dijkstra's calculus. *ACM Transactions on Programming Languages and Systems*, 11(4):517–561, October 1989.

[55] J.E. Nicholls and Sørensen I.H. Collaborative project in software development. IBM Hursley Park and Programming Research Group Oxford.

[56] T. Nipkow. Non-deterministic data types. *Acta Informatica*, 22:629–661, 1986.

[57] J.C. Reynolds. *The Craft of Programming*. Prentice-Hall, London, 1981.

[58] K.A. Robinson. From specifications to programs. Department of Computer Science, University of New South Wales, Australia.

[59] J.M. Spivey. *Understanding Z: a Specification Language and its Formal Semantics*. Cambridge University Press, 1988.

[60] J.M. Spivey. *The Z Notation: A Reference Manual*. Prentice-Hall, London, 1989.

[61] J.E. Stoy. *Denotational Semantics: the Scott-Strachey Approach to Programming Language Theory*. MIT Press, 1977.

[62] N. Wirth. *Programming in Modula-2*. Springer, 1982.

Authors' addresses

Carroll Morgan
Programming Research Group, 8–11 Keble Road, Oxford OX1 3QD, U.K.
carroll@prg.ox.ac.uk

Paul Gardiner
Programming Research Group, 8–11 Keble Road, Oxford OX1 3QD, U.K.
paulg@prg.ox.ac.uk

Ken Robinson
Department of Computer Science, University of New South Wales,
P.O. Box 1, Kensington 2033, Australia.
kenr@spectrum.cs.unsw.oz.au

Trevor Vickers
Department of Computer Science, Australian National University,
P.O. Box 4, Canberra 2601, Australia.
trev@cs.anu.edu.au

and

Ralph Back
Department of Computer Science, Åbo Akademi,
Lemminkäinengatan 14, SF–20500 Turku, Finland.
backrj@aton.abo.fi

Joe Morris
Department of Computing, Glasgow University, Glasgow G12 8QQ, U.K.
jmm@dcs.glasgow.ac.uk